Advances in Experimental Medicine and Biology

VOLUME 853

Advances in Experimental Medicine and Biology presents multidisciplinary and dynamic findings in the broad fields of experimental medicine and biology. The wide variety in topics it presents offers readers multiple perspectives on a variety of disciplines including neuroscience, microbiology, immunology, biochemistry, biomedical engineering and cancer research. *Advances in Experimental Medicine and Biology* has been publishing exceptional works in the field for over 30 years and is indexed in Medline, Scopus, EMBASE, BIOSIS, Biological Abstracts, CSA, Biological Sciences and Living Resources (ASFA-1), and Biological Sciences. The series also provides scientists with up to date information on emerging topics and techniques.

2013 Impact Factor: 2.012

More information about this series at http://www.springer.com/series/5584

Moneeb Ehtesham
Editor

Stem Cell Biology in Neoplasms of the Central Nervous System

Springer

Editor
Moneeb Ehtesham
Department of Neurosurgery
Vanderbilt University Medical Center
Nashville, TN, USA

ISSN 0065-2598 ISSN 2214-8019 (electronic)
Advances in Experimental Medicine and Biology
ISBN 978-3-319-16536-3 ISBN 978-3-319-16537-0 (eBook)
DOI 10.1007/978-3-319-16537-0

Library of Congress Control Number: 2015937184

Springer Cham Heidelberg New York Dordrecht London

Printed on acid-free paper

Springer International Publishing AG Switzerland is part of Springer Science+Business Media (www.springer.com)

To my juniors—J.A.M., A.M.M., S.L.Z., A.S.
No man could ask for a better team
in the heat of battle

Acknowledgements

A volume like this is beyond the scope of one person/editor's effort. There are countless hours spent contacting authors, reviewing submissions, and cross-checking references and publication permissions—tedious work that can be draining but remains essential for a project like this to succeed. This project would have failed miserably if it were not for the tireless effort and dedication of Imad Saeed Khan, M.D., whose assistance in making this volume a reality is tremendously appreciated by me and all those involved. Thank you, Imad, for your passion in helping make this idea a reality. I would also like to thank Fiona Sarne, Joy Evangeline Bramble, and Sabina Ashbaugh at Springer who, together, were always available to assist and shepherd this project through, and whose help made this volume a pleasure to work on.

Contents

Contributors

David Bakhshinyan McMaster Stem Cell and Cancer Research Institute, McMaster University, Hamilton, ON, Canada

Department of Biochemistry and Biomedical Sciences, Faculty of Health Sciences, McMaster University, Hamilton, ON, Canada

Christoph Patrick Beier Department of Neurology, University of Southern Denmark, Odense C, Denmark

Institute of Clinical Research, University of Southern Denmark, Odense M, Denmark

Dagmar Beier Department of Neurology, University of Southern Denmark, Odense C, Denmark

Institute of Clinical Research, University of Southern Denmark, Odense M, Denmark

Vivian Capilla-Gonzalez Stem Cells Department, Andalusian Molecular Biology and Regenerative Medicine Centre, Seville, Spain

Department of Neurosurgery and Oncology, Johns Hopkins University School of Medicine, Baltimore, MD, USA

Lincoln Edwards Maxine Dunitz Neurosurgical Institute, Cedars-Sinai Medical Center, Los Angeles, CA, USA

Moneeb Ehtesham Department of Neurological Surgery, Vanderbilt University Medical Center, Nashville, TN, USA

Sigurd Fosmark Department of Pathology, Odense University Hospital, Odense C, Denmark

Institute of Clinical Research, University of Southern Denmark, Odense M, Denmark

Neha Garg McMaster Stem Cell and Cancer Research Institute, McMaster University, Hamilton, ON, Canada

Hugo Guerrero-Cazares Department of Neurosurgery, Johns Hopkins University School, Baltimore, MD, USA

Sofie Hellwege Department of Pathology, Odense University Hospital, Odense C, Denmark

Institute of Clinical Research, University of Southern Denmark, Odense M, Denmark

Imad Saeed Khan Section of Neurosurgery, Geisel School of Medicine at Dartmouth, Lebanon, NH, USA

Bjarne Winther Kristensen Department of Pathology, Odense University Hospital, Odense C, Denmark

Institute of Clinical Research, University of Southern Denmark, Odense M, Denmark

Yael Kusne Barrow Brain Tumor Research Center, Phoenix, AZ, USA

Emily Lavell Department of Neurosurgery and Oncology, Johns Hopkins University School of Medicine, Baltimore, MD, USA

Mecca Madany Maxine Dunitz Neurosurgical Institute, Cedars-Sinai Medical Center, Los Angeles, CA, USA

Graduate Program in Biological Science & Translational Medicine, Cedars-Sinai Medical Center, Los Angeles, CA, USA

Branavan Manoranjan McMaster Stem Cell and Cancer Research Institute, McMaster University, Hamilton, ON, Canada

Michael G. DeGroote School of Medicine, McMaster University, Hamilton, ON, Canada

Department of Biochemistry and Biomedical Sciences, Faculty of Health Sciences, McMaster University, Hamilton, ON, Canada

Alfredo Quiñones-Hinojosa Department of Neurosurgery and Oncology, Johns Hopkins University School of Medicine, Baltimore, MD, USA

Anthony E. Rizzo Department of Stem Cell Biology and Regenerative Medicine, Lerner Research Institute, Cleveland Clinic, Cleveland, OH, USA

Nader Sanai Barrow Brain Tumor Research Center, Phoenix, AZ, USA

Sheila K. Singh McMaster Stem Cell and Cancer Research Institute, McMaster University, Hamilton, ON, Canada

Michael G. DeGroote School of Medicine, McMaster University, Hamilton, ON, Canada

Department of Biochemistry and Biomedical Sciences, Faculty of Health Sciences, McMaster University, Hamilton, ON, Canada

Department of Surgery, Faculty of Health Sciences, McMaster University, Hamilton, ON, Canada

Mia Dahl Sørensen Department of Pathology, Odense University Hospital, Odense C, Denmark

Institute of Clinical Research, University of Southern Denmark, Odense M, Denmark

Tom M. Thomas Maxine Dunitz Neurosurgical Institute, Cedars-Sinai Medical Center, Los Angeles, CA, USA

Graduate Program in Biological Science & Translational Medicine, Cedars-Sinai Medical Center, Los Angeles, CA, USA

Jennifer S. Yu Department of Radiation Oncology, Cleveland Clinic, Cleveland, OH, USA

Department of Stem Cell Biology and Regenerative Medicine, Lerner Research Institute, Cleveland Clinic, Cleveland, OH, USA

John S. Yu Maxine Dunitz Neurosurgical Institute, Cedars-Sinai Medical Center, Los Angeles, CA, USA

Regulation of Subventricular Zone-Derived Cells Migration in the Adult Brain

Vivian Capilla-Gonzalez, Emily Lavell, Alfredo Quiñones-Hinojosa, and Hugo Guerrero-Cazares

Abstract The subventricular zone of the lateral ventricles (SVZ) is the largest source of neural stem cells (NSCs) in the adult mammalian brain. Newly generated neuroblasts from the SVZ form cellular chains that migrate through the rostral migratory stream (RMS) into the olfactory bulb (OB), where they become mature neurons. Migration through the RMS is a highly regulated process of intrinsic and extrinsic factors, orchestrated to achieve direction and integration of neuroblasts into OB circuitry. These factors include internal cytoskeletal and volume regulators, extracellular matrix proteins, and chemoattractant and chemorepellent proteins. All these molecules direct the cells away from the SVZ, through the RMS, and into the OB guaranteeing their correct integration. Following brain injury, some neuroblasts escape the RMS and migrate into the lesion site to participate in regeneration, a phenomenon that is also observed with brain tumors. This review focuses on factors that regulate the migration of SVZ precursor cells in the healthy and pathologic brain. A better understanding of the factors that control the movement of newly generated cells may be crucial for improving the use of NSC-replacement therapy for specific neurological diseases.

Keywords Neural stem cells • Subventricular zone • Rostral migratory stream • Neuroblasts • Neuronal migration • Regulation of migration • Adult neurogenesis • Brain tumors

---

V. Capilla-Gonzalez
Stem Cells Department, Andalusian Molecular Biology and Regenerative Medicine Centre, Seville, Spain

Department of Neurosurgery and Oncology, Johns Hopkins University School of Medicine, Baltimore, MD, USA

E. Lavell • A. Quiñones-Hinojosa
Department of Neurosurgery and Oncology, Johns Hopkins University School of Medicine, Baltimore, MD, USA

H. Guerrero-Cazares, M.D., Ph.D. (✉)
Department of Neurosurgery, Johns Hopkins University,
1550 Orleans St, CRBII, Room 247B, Baltimore, MD 21201, USA
e-mail: hugo@jhmi.edu

M. Ehtesham (ed.), *Stem Cell Biology in Neoplasms of the Central Nervous System*, Advances in Experimental Medicine and Biology 853,
DOI 10.1007/978-3-319-16537-0_1

Introduction

Adult neurogenesis mainly occurs in two regions of the mammalian brain, the subgranular zone of the hippocampus and the subventricular zone (SVZ) of the lateral ventricles [1]. In rodents' adult SVZ, highly migratory neuroblasts arise from neural stem cells (NSCs). These neuroblasts move tangentially through a specialized path called the rostral migratory stream (RMS) to reach their final destination, the olfactory bulb (OB) [2–10]. Once in the OB, SVZ-derived neuroblasts incorporate into the OB circuitry and differentiate into interneurons [6, 11–14]. The existence of a similar migratory pathway in the adult human brain remains highly controversial. However, there is prominent neuroblast migration from the ventricular walls into the olfactory tract in the human fetal brain that appears to decrease with development [15–22].

The regulation of SVZ-derived cells' migration in the adult brain involves multiple processes including dynamic cell–cell communication, cell–extracellular matrix interactions, as well as chemo repellent and chemo attractant signals [23–26]. Under certain pathological conditions, these processes are modified to redirect the migration of SVZ-derived cells and provide support in damaged areas [26–31]. We review the roles of the different mechanisms that regulate the migration of newly generated cells from the SVZ into the OB, as well as their migration into damaged brain areas and brain tumors. A more complete understanding of the mechanisms regulating this migratory capacity could provide valuable insight in the possible use of NSCs in treatment of neurological disorders.

Adult Neurogenic Niche: The Subventricular Zone

The SVZ is the major source of NSCs in the adult brain of mammals, including humans. In this region, NSCs have been identified as a subpopulation of astroglial cells that are able to differentiate into any of the main cell types of the central nervous system, i.e., neurons, oligodendrocytes, or astrocytes [32, 33]. These cells organize in the SVZ to confer a unique cytoarchitecture, which presents remarkable differences between rodents and humans [21].

Rodents

The SVZ is found behind a layer of ependymal cells (type E cells) that separates it from the ventricle cavity. The NSCs within the rodent SVZ correspond to a subpopulation of astrocytes called B1 cells. These cells display a primary cilium in the apical surface that extends into the ventricle cavity, which has been related to the proliferative activity of B1 cells [34–39]. Typically, B1 astrocytes divide asymmetrically, maintaining their population and giving rise to the highly proliferative, transient amplifying progenitors (type C cells), which then become migratory

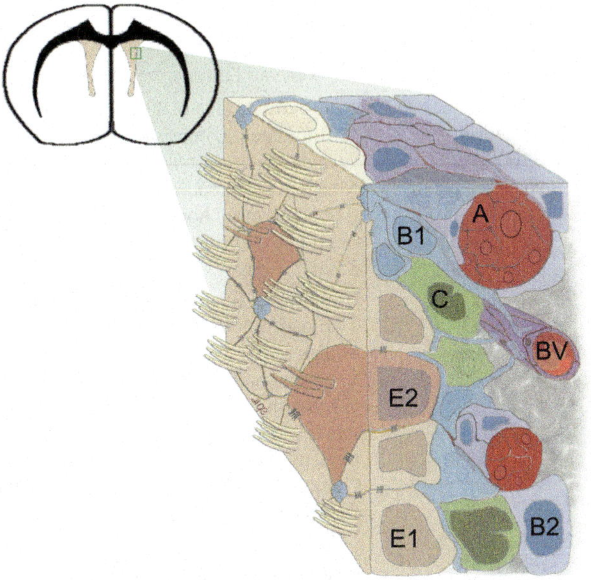

Fig. 1 Cytoarchitecture of the subventricular zone (SVZ) in rodents. Biciliated (E2) and multiciliated (E1) cells form a monolayer on the ventricular wall. E1 and E2 cells are organized in a pinwheel fashion surrounding B1 cell cilia. B1 cells extend one apical process that maintains contact with the ventricle and another process that contacts blood vessels. Type B2 cells are not in contact with the ventricular wall and surround the migratory A cells. Type C cells are highly proliferative and separated from the E1 and E2 cells by astrocytic processes. Type A cells organize in chains to migrate tangentially while surrounded by astrocytic cells

neuroblasts (type A cells). Additionally, the SVZ contains a subpopulation of non-neurogenic astrocytes (B2 astrocytes) that are located at the underlying striatal parenchyma and do not make contact with the ventricle. These cells function as support for neuroblasts migration toward the OB (Fig. 1) [2–6].

All SVZ cell types can be distinguished from each other by their ultrastructural characteristics and specific molecular markers (Table 1). Briefly, B1 astrocytes have light cytoplasm with abundant intermediate filaments, and express glial fibrillary acidic protein (GFAP), nestin and vimentin. Type C cells are large cells with dark cytoplasm and many mitochondria. This cell type expresses the molecular markers Mash1, DLX-2, and nestin. Neuroblasts (Type A cells) are small, elongated cells that typically express nestin, Doublecortin (Dcx), polysialylated-neural cell adhesion molecule (PSA-NCAM), and Neuron-specific class III beta-tubulin (Tuj1). In the SVZ, neuroblasts form large chains that are surrounded by astrocytes. The B2 astrocytes are similar to B1 astrocytes but have higher number of intermediate filaments and do not contact the ventricle. Type E cells exhibit multiple cilia and microvilli on their apical surface and express markers such as vimentin, nestin, S100, and CD24 [10, 35, 40–43].

As neuroblasts migrate away from the SVZ, they form large cellular chains, ensheathed by processes of B2 astrocytes (Fig. 2a, b). These structures constitute

Table 1 Ventricular zone cell types and characteristics

Cell type	Molecular markers	Microscopic and ultrastructural characteristics
E1 Multiciliated ependymal cells	CD24 Vimentin S100 beta Nestin	– Approximate size of 115 μm^2 – Multiciliated, comprise most of the ventricular surface – Multiple basal bodies – Abundant mitochondria concentrated around the basal bodies – Small Golgi
E2 Biciliated Ependymal cells	CD24 Vimentin S100 beta Nestin	– Comprise 5 % of ventricular surface – Two partially invaginated cilia (9+2) and two basal bodies – Spherical nuclei with no invaginations and dispersed chromatin – Abundant mitochondria concentrated around the nucleus – Complex set of electron-dense particles surrounding the basal body – Small Golgi
B1 Astrocytes	GFAP Nestin Cd133 Vimentin	– Approximate size of 58 μm^2 – Have a long basal process that terminates on blood vessels and an apical ending at the ventricle cavity – Reside adjacent to the ependymal layer – Light cytoplasm – Single, small cilia that contact the ventricle – Basal body – Non-clumped chromatin – Medium Golgi – Intermediate filaments – Few free ribosomes
B2 Astrocytes	GFAP Vimentin Nestin	– Approximate size of 43 μm^2 – Are located at the underlying striatal parenchyma – Surround neuroblasts chains – Single, small cilia that do not contact the ventricle – Clumped chromatin – Medium Golgi – High number of intermediate filaments – Few free ribosomes
C Transient amplifying cells	Nestin Mash1 DLX-1	– Approximate size of 62 μm^2 – Found throughout lateral wall of lateral ventricle but not in RMS – Actively proliferating – Large, dark, and undifferentiated cytoplasm
A Neuroblasts	Nestin Dcx PSA-NCAM Tuj1 Ki-67 Pax-6 NKCC1 Robo	– Small size, approximately 31 μm^2 – Organize in chains as migrating through the RMS – Small, elongated, and dark cytoplasm

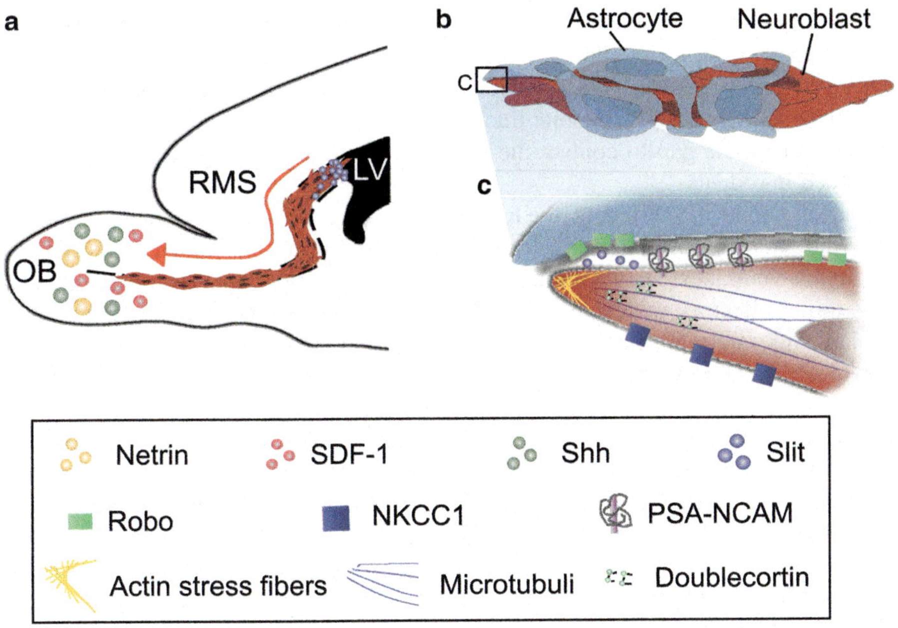

Fig. 2 Neuroblasts migration from the SVZ and through the RMS in rodents. (**a**) This process is controlled by extrinsic chemoattractant and chemorepellent factors. (**b**) Migratory neuroblasts form cellular chains surrounded by astrocytic gliotubes. (**c**) At the cellular level, membrane, cytoplasmatic and cytoskeletal proteins regulate cell–cell and cell–microenvironment interactions to mediate cell migration

the gliotubes that SVZ-derived cells use to migrate toward the OB. Gliotubes of astrocytes converge in the anterior dorsal horn of the SVZ and initiate the RMS, which extends rostrally into the central region of the OB, where neuroblasts become mature neurons [7, 10, 32, 44–49]. Thus, neuroblasts migrate up to 5 mm before reaching their final destination in the mouse brain [10].

In the rodent brain, the RMS appears during the embryonic development, on E15–17, and is maintained into adulthood [10, 50]. However, the number of neuroblasts that migrate toward the OB decreases with aging and the RMS tends to disappear in aged mice. Even so, the cytoarchitecture of the migratory route remains similar during its development [9, 51].

Humans

In the adult human SVZ, the presence of GFAP-positive NSCs has also been described in detail [21]. However, the SVZ cytoarchitecture is significantly different from that in rodents. The adult human SVZ is composed largely of astrocytes expressing GFAP and ependymal cells, while there has been no description of type

C cells or chains of migratory neuroblasts [19, 21]. In contrast to rodents, human SVZ astrocytes are accumulated in a ribbon that is not adjacent to the ependymal layer. Instead, a gap that is largely devoid of cells separates the astrocytic ribbon from the ependymal cells. Despite this, some astrocytes are found to extend a long process across the gap to contact the ventricular surface, similar to B1 astrocytes described in rodents.

During human development, the RMS has been described in the fetal brain where a rostral extension of neuroblasts is evident [20]. During the second trimester of gestation a ventral extension of the anterior horn of the lateral ventricle connects the SVZ to the olfactory peduncle. At this stage, Dcx positive cells are abundant and display a migratory morphology in which they are mixed with GFAP-positive cells that are part of reminiscent gliotubes [20]. After birth, an extensive corridor of migrating immature neurons remains in the RMS of infants (before 18 months of age), but declines in older children (7 years of age) [22]. Currently, the existence of a real RMS in the adult human brain is a subject of debate [15–19]. It has been suggested that a human RMS arises from the adult SVZ adjacent to the lateral ventricle overlying the caudate nucleus to the olfactory tract in the base of the brain. Here, neuroblasts do not form chains but exhibit migratory morphologies and co-express Dcx and PSA-NCAM [17]. This RMS pathway presents an approximate total length of 17 mm and is organized around a lateral ventricular extension that ends in the OB [16, 18]. Conversely, other studies have not been able to confirm a true ventricular extension in the adult human brain and have suggested that the RMS tends to disappear after birth, as is nearly extinct by adulthood [15, 21, 22]. This phenomenon would be similar to that described in rodents during aging [9], but occurring in an earlier stage of development in humans [15–19, 21].

Regulation of Neuroblasts Migration Toward the RMS

As neuroblasts are formed in the SVZ and begin to migrate, they organize into defined chain-like structures surrounded by GFAP positive cells called gliotubes (Fig. 2b). The process of neuroblasts migration is simultaneously regulated by a complex interplay of intrinsic and extrinsic factors. Multiple proteins are involved in controlling the mechanisms that determine the direction, speed, and morphology of neuroblasts through cell–cell and cell–extracellular matrix interactions as well as chemoattractive and chemorepellent signaling [24, 52–57] (Fig. 2c).

Morphological Changes in Migrating Cells

Doublecortin

Dcx is a microtubule-associated, stabilizing protein that is expressed in neural precursor cells while actively dividing. It continues to be expressed as neuroblasts migrate and begin differentiation into neurons within the developing and adult mouse and human brains [58]. Dcx plays a central role in the regulation of microtubule dynamics and stability throughout morphogenesis. Mutant alleles of Dcx are associated with impairments in proper cell migration during development and contribute to diseases such as lissencephaly in humans [59]. Immunostaining reveals the highest expression levels of Dcx within the soma and leading processes of migrating cells as well as in distal ends of developing processes in dividing cells [40, 60]. In vitro knockdown of Dcx in SVZ-derived cells of adult mice caused a significant decrease in migratory capacity [61]. In addition, Dcx deletion in vivo results in branching defects in the dynamics of migrating cells, leading to the formation and division of growth cones. This causes Dcx knockout cells to produce more, but less stable processes. The combination of these deficits leads to decreased tangential migration and an accumulation of neuroblasts, which results in a thickened SVZ and disorganized RMS [62, 63]. These results suggest Dcx plays an important role in migration through the maintenance of bipolar morphology during the migration of neuroblasts.

NKCC1

Cell migration is accompanied by cell volume changes in the mammalian brain as morphology of a migrating cell changes from a round, resting state to one that is polarized [55, 56]. Ion transporters, specifically those involved in volume regulation, adopt a polarized distribution in migrating cells. Therefore, it is hypothesized that inhibition of these cell volume regulators impairs cell migration [57, 64]. The ionic cotransporter of Na^+-K^+-$2Cl^-$ (NKCC1) is a fundamental transporter utilized in the regulation of intracellular volume and in the accumulation of intracellular Cl^- [65–67]. NKCC1 mediates the movement of Na^+, K^+, and Cl^- ions across the plasma membrane using energy stored in the Na^+ gradient, generated by the Na^+/K^+ ATPase. This movement of ions with osmotically obligated water across plasma membranes plays an important role in the maintenance of volume within the cell [68, 69]. In the RMS, NKCC1 activity is necessary to conserve normal migration speed of neuroblasts by maintaining elevated intracellular concentration of Cl^- in migrating neuroblasts [54]. Interestingly, the role of NKCC1 in cell migration is also associated with cytoskeleton rearrangements using in vitro assays with glioma cells [70], which could also be happening in the migrating neuroblasts.

Cdk5

Cyclin-Dependent Kinase 5 (Cdk5) modulates the migration of prenatal and postnatal neural cells through phosphorylation of cytoskeletal proteins, including neurofilament proteins and microtubule associated protein tau [71–74]. Cdk5 plays an important role in migration of neuroblasts, and in Cdk5-deficient mice, neuroblasts fail to form properly oriented chains from the SVZ toward the OB. Cdk5 deletion is associated with impairments in chain formation, speed, directionality, and extension of leading processes of neuroblasts [74]. Consequently, Cdk5-deficient mice present an accumulation of neuroblasts in the RMS and SVZ due to a reduced cell migration.

Cell Adhesion Proteins

PSA-NCAM

The PSA-NCAM is a glycoform of the NCAM immunoglobulin family that presents anti-adhesive properties and is considered a hallmark of neurons capable of dynamic change and of migrating neurons, particularly those within the glial tubes of the RMS [7, 45, 75, 76]. The specific role of PSA-NCAM on cell migration focuses on providing a favorable extracellular environment on which the neuroblasts may move. This protein increases the intercellular space between the glial tube and migrating neurons to allow room for selective encounter of complementary receptors between cells and allow dynamic changes required for movement [77–79]. In addition, further studies demonstrate that this adhesion molecule is important for interactions between SVZ cells and suggests that SVZ cells use each other as their migratory substrate [78]. Deficits of PSA-NCAM lead to impairments of neuroblasts migration away from the SVZ, which then results in a reduction in OB size [80–82].

Integrins

Integrins are heterodimeric transmembrane receptors that mediate adhesion between adjacent cells by communicating information between the external matrix and internal cytoskeleton of the cell. This receptor also recognizes multiple extracellular ligands such as collagen, fibronectin, tenascin-C, laminin, and intercellular adhesion molecule 1 (ICAM-1) [83]. Of these ligands, laminin is expressed in the RMS of rats during E16–P4, when migrating cells are known to express integrins [84, 85]. Studies of interactions between the integrins and laminin extracellular matrix have revealed that neuroblasts use the interaction to regulate production of protrusions and in tangential migration [84–86]. In accordance, the blockade of endogenous $\alpha 6$ or $\beta 1$ integrin subunit binding to its receptor laminin disrupted the cohesive nature of the RMS, while the injection of an exogenous laminin peptide redirected the neuroblasts away from the RMS and towards the site of infusion [85]. Moreover, it was proposed that the integrins play a key role in the formation of protrusions and cellular translocation, with different integrins participating at different developmental stages [84]. In addition, it has been demonstrated that laminin can be localized to cell chains in the RMS and that the formation of migratory chains of neuroblasts are in part controlled by $\beta 1$ integrin interaction with laminin [11].

Chemorepellent and Chemoattractant Signals

Netrins

Netrins are a family of diffusible chemotropic proteins that regulate axon guidance and cell migration during development [87]. In the RMS, a chemo gradient of netrin-1 is created by mitral cells in the OB. Netrin is recognized by migrating neuroblasts which express neogenin and deleted in colorectal carcinoma (DCC) receptors for netrin-1 [84]. The interaction of netrin-1 and DCC is suggested to play a role in directing migration of neuroblasts from the SVZ toward the OB by regulating the formation and direction of what is usually a single, prominent and long-lived polarized cellular protrusion. Although netrin-1 expression is greatly reduced in postnatal stages, NSCs population of the adult forebrain preserves neogenin expression [83]. This finding suggests that other netrins could be interacting with neogenin to regulate cell migration during adulthood. For instance, netrin-4 is expressed by a subpopulation of astrocytes within the RMS and the OB of adult mice [88]. Although a binding between netrins-4 and neogenin has not been demonstrated, netrin-4 interacts with laminin γ1 chain and the α6β1 integrin to form a protein complex on the surface of NSCs. This complex promotes proliferation, adhesion, and migration through an α6β1 integrin-mediated signaling pathway in vitro [88].

Sonic Hedgehog

Sonic hedgehog (Shh) is a secreted protein that plays complex and crucial roles during organ formation in the developing embryo by orchestrating reciprocal communication between different cells and tissues. This protein regulates a variety of processes such as proliferation, differentiation and specification of cell fates, especially within the nervous system [89]. In the mature adult brain, Shh continues to have importance and acts as chemoattractant in regulating the migration of neuroblasts along the RMS and is associated with the genesis of specific neuronal progeny within the OB [90–92]. The blockage of Shh can stop the migration of SVZ-derived neuroblasts, while the exogenous presence of Shh in other brain regions leads to the deviation of the neuroblasts from their normal migratory pathway, making this a highly influential factor for neural progenitor cells [90].

Slit-Robo Signaling

Slit and Robo form part of a chemorepulsive system which is related to various migratory processes including axon guidance and branching, NSC migration, leukocyte chemotaxis, angiogenesis, and glioma cell migration [93, 94]. In vertebrates, Robo proteins constitute a family of proteins with four homologs (Robo1–Robo4) that act as receptors for Slits, which have three homologs (Slit1–Slit3). When Slit binds Robo, the cytoplasmic conserved (CC) sequences of the receptor interact with different intracellular effectors to control cell motility [93, 95]. In the RMS, Slit1 and Slit2 are expressed in the embryonic and adult septum. Slit proteins repel OB-bound migrating neuroblasts away from the embryonic and postnatal SVZ. Consequently, Slit1/2 KO mice present a disrupted RMS and significantly smaller OB when compared to wild type mice [96]. Furthermore, neuroblasts open their own pathway through the brain parenchyma by releasing Slit1, which is recognized by Robo-expressing astrocytes. This interaction induces a chemorepellent response on astrocytes, which opens a tube through which neuroblasts may migrate [23, 24].

SDF-1

Stromal cell-derived factor 1 (SDF-1) is a chemoattractant for T cell lymphocytes upon binding to its receptors CXCR4 and RDC1/CXCR7. CXCR4 and CXCR7 expression in rats is high in proliferative areas such as the OB of the brain during embryonic development and is also expressed in the developing mouse brain until maturity. At all stages, the receptor is transcribed in ventricular zones of neuronal proliferation. SDF-1 is involved in a wide array of developmental changes such as neuronal migration, axon guidance, and axon elongation [97, 98]. While postnatal proliferative cells in mouse brain have been demonstrated to co-express the CXCR4 and CXCR7 receptors, only SVZ cells expressing CXCR7 are found to migrate through the RMS. Alterations in the expression of CXCR7 lead to disruption in RMS formation. It is unclear whether CXCR7 works as a separate receptor on its own or acts as a regulator of the functions of SDF-1 binding to CXCR4 [99, 100].

Migration in Response to Brain Injury

In addition to supporting constitutive neurogenesis into the OB, SVZ-derived cells are able to ectopically migrate to damaged areas and participate in brain regeneration by activating repair mechanisms (Fig. 3) [101–106]. For instance, after a stroke or a demyelinating damage, new cells are generated and mobilized from the SVZ to the site of the injury, where they differentiate into neurons or myelinating oligodendrocytes, respectively [27–29, 31, 102, 104, 107–110]. The proliferative and migratory responses of SVZ progenitors is mainly due to the production of inflammatory cytokines or diffusible growth factors in the lesion site, which strongly regulate the function of NSCs and attract the newborn cells into the injury [111–115].

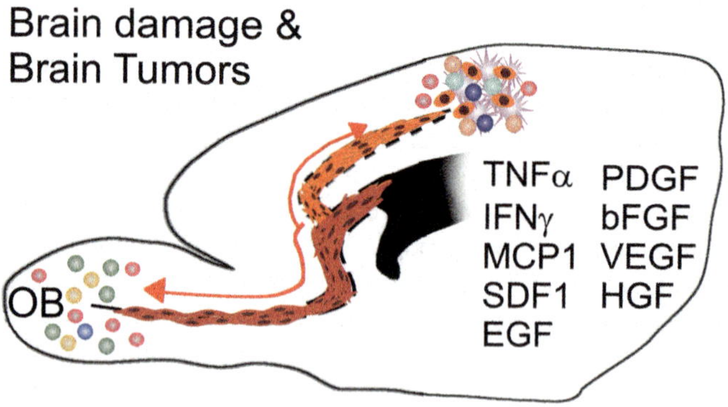

Fig. 3 Cellular migration from the SVZ in response to a brain damage or brain tumor. Inflammatory chemokines and growth factors released by an injured site or brain tumor redirect the migration of SVZ-derived cells

Cytokines

Brain damages are frequently accompanied by neuroinflammatory processes, which lead to the release of numerous pro-inflammatory cytokines by local microglia and other immune cells. Cytokines are small signaling molecules that bind to specific receptors, expressed by NSC, and initiate different cascades of intracellular signaling [116, 117]. Thus, when a brain damage occurs, cytokines mediate the recruitment of migratory cells into the injury [118, 119]. In this context, the tumor necrosis factor-alpha (TNFα) and interferon-gamma (IFNγ) play key roles in the recruitment of new cells into the area of neuroinflammation [120]. However, cytokines activity frequently promotes the production of chemokines, a small family of cytokines. Thus, TNFα increases the expression of the monocyte chemotactic protein-1 (MCP-1), also known as chemokine (C-C motif) ligand 2 (CCL2) and small inducible cytokine A2. After brain damage, MCP-1 is involved in increasing the migration of SVZ-derived cells and promoting neuronal and oligodendrocytic differentiation [121–125]. After a cortical lesion, the SVZ increases its cell proliferation and neurogenesis, followed by a subsequent ectopic migration of neural progenitors into the lesion site that is regulated by the stromal-derived-factor-1 (SDF-1) [116, 126–129]. These progenitors differentiate into glial cells and, to a lesser extent, into neurons [126, 127].

Other Factors Regulating Migration in the Damaged Brain

Growth factors are also involved in the migration of precursor cells into a lesion site. For instance, in demyelinating disease, epidermal growth factor (EGF), basic fibroblast growth factor (bFGF), and platelet-derived growth factor (PDGF) mediate the migration of oligodendroglial precursors into the a lesion site [109, 130–132]. In an ischemic brain, the vascular endothelial growth factor (VEGF) is known to assist the mobilization of the newborn cells from the SVZ through the infarct area [133–135].

Endogenous matrix metalloproteinase (MMP) has also been involved in promoting differentiation and migration of NSCs after brain damage [136]. In this study, NSCs differentiated into migratory cells in response to stroke-induced chemokines (i.e., SDF-1 and vascular endothelial growth factor) and expressed higher levels of MMP-3 and MMP-9 than non-migratory cells. Thus, MMPs were suggested to play an important role for mediating the NSCs response to extrinsic signals [30, 136]. Similarly, reelin, a secreted glycoprotein found in the extracellular matrix, is upregulated in lesion sites and induces the escape of neuroblasts from the RMS and attraction to the damaged area [137].

Stem Cell Tropism for Brain Tumors

Similar to brain damage, brain tumors exert a tropic effect on NSCs that attracts them to migrate and associate with infiltrative cancer tissue [138]. Gliomas are the most common intra-axial brain tumors, including glioblastoma (GBM) the most common and aggressive glioma in adults [139]. Their extensive invasive capacity makes total resection an almost impossible task, which leads to high rates of recurrence and short survival despite the use of current therapeutic strategies [140]. When NSCs are injected into experimental gliomas in mice, they distribute themselves throughout the tumor and are found migrating alongside migratory tumor cells. In addition, when NSCs are implanted in healthy tissue away from the tumor or intravascularly outside the central nervous system, NSCs are able to migrate through normal tissue in order to target tumor cells [141].

As in the brain damage context, stem cells' ability to migrate towards cancerous tissue is controlled by a complex interplay of many factors, including diffusible signals as well as extracellular matrix (ECM) signals. Under hypoxic conditions, gliomas are known to increase production of SDF-1, uPA, VEGF, and hepatocyte growth factor. Remarkably, when NSCs are exposed to a hypoxic environment, they upregulate the receptors to each of these signals: CXCR4, uPAR, VEGF2, and c-Met [142, 143]. These signaling pathways are all critical for allowing NSCs to maintain their tracking ability and navigate their way to a tumor mass as well as follow migratory cancer cells as they invade surrounding tissue. Some of these pathways converge on PI3K to influence cytoskeletal rearrangement and movement [144].

In addition to secreted signals, migratory glioma cells modify the ECM as they invade the normal brain parenchyma. These modifications leave a specifically altered trail of molecules that serves as a path for migratory neural progenitor cells. Such molecules include tenascin, fibronectin, laminin, vitronectin, and different types of collagen are components of this path [143].

A potential use of these properties is to utilize NSCs as vehicles to deliver therapeutic proteins. Due to the genomic stability of NSCs, therapeutic genes can be inserted into them, and will continue to be expressed as the cells reach cancerous tissue. This approach has been experimentally tested using prodrug-activating enzymes (cytosine deaminase, carboxylesterase, thymidine kinase), interleukins (IL-2, IL-4, IL-12, IL-23), interferon-b, apoptosis-promoting genes (tumor necrosis factor-related apoptosis-inducing ligand) and metalloproteinases (PEX) [141, 145, 146]. Another use of these cells as delivery vehicles is to help monoclonal antibodies penetrate dense, hypoxic tumor masses [145]. Stem cell delivery of antibody treatment holds promise to overcome these issues and provide a very specific, penetrative treatment with significantly less damage to healthy tissue.

Conclusions

Newborn neuroblasts in the SVZ move tangentially long distances forming migratory chains along the RMS toward the OB. However, in the presence of a brain injury, SVZ-derived cells can modify their traditional migratory route to reach the damage and participate in tissue repair. While the normal migration is mainly controlled by transmembrane and extracellular matrix proteins, migration in response to brain damages is regulated by chemokines and cytokines. In humans, these migratory behaviors decrease with age and are apparently absent in the adult. However, the presence of NSCs in the adult human brain suggests that these cells have the potential to originate migratory cells that could respond to brain tissue damage. Furthermore, the use of NSCs in cell replacement therapy continues to be investigated. Controlling the factors and mechanisms involved in modulating the migration of new cells is a potential tool for cell therapy.

References

1. Johansson CB, Momma S, Clarke DL, Risling M, Lendahl U, Frisen J. Identification of a neural stem cell in the adult mammalian central nervous system. Cell. 1999;96(1):25–34. doi:10.1016/s0092-8674(00)80956-3.
2. Doetsch F, Garcia-Verdugo JM, Alvarez-Buylla A. Regeneration of a germinal layer in the adult mammalian brain. Proc Natl Acad Sci U S A. 1999;96(20):11619–24.
3. Ponti G, Obernier K, Guinto C, Jose L, Bonfanti L, Alvarez-Buylla A. Cell cycle and lineage progression of neural progenitors in the ventricular-subventricular zones of adult mice. Proc Natl Acad Sci U S A. 2013;110(11):E1045–54. doi:10.1073/pnas.1219563110. 1219563110 [pii].
4. Luskin MB, Zigova T, Soteres BJ, Stewart RR. Neuronal progenitor cells derived from the anterior subventricular zone of the neonatal rat forebrain continue to proliferate in vitro and express a neuronal phenotype. Mol Cell Neurosci. 1997;8(5):351–66. doi:10.1006/mcne.1996.0592. S1044-7431(96)90592-8 [pii].
5. Lazarini F, Lledo PM. Is adult neurogenesis essential for olfaction? Trends Neurosci. 2011;34(1):20–30. doi:10.1016/j.tins.2010.09.006. S0166-2236(10)00137-2 [pii].
6. Imayoshi I, Sakamoto M, Ohtsuka T, Takao K, Miyakawa T, Yamaguchi M, Mori K, Ikeda T, Itohara S, Kageyama R. Roles of continuous neurogenesis in the structural and functional integrity of the adult forebrain. Nat Neurosci. 2008;11(10):1153–61.
7. Lois C, Garcia-Verdugo JM, Alvarez-Buylla A. Chain migration of neuronal precursors. Science. 1996;271(5251):978–81.
8. Carleton A, Petreanu LT, Lansford R, Alvarez-Buylla A, Lledo PM. Becoming a new neuron in the adult olfactory bulb. Nat Neurosci. 2003;6(5):507–18. doi:10.1038/nn1048.
9. Capilla-Gonzalez V, Cebrian-Silla A, Guerrero-Cazares H, Garcia-Verdugo JM, Quinones-Hinojosa A. The generation of oligodendroglial cells is preserved in the rostral migratory stream during aging. Front Cell Neurosci. 2013;7:147. doi:10.3389/fncel.2013.00147.
10. Lois C, Alvarez-Buylla A. Long-distance neuronal migration in the adult mammalian brain. Science. 1994;264(5162):1145–8.
11. Belvindrah R, Lazarini F, Lledo PM. Postnatal neurogenesis: from neuroblast migration to neuronal integration. Rev Neurosci. 2009;20(5–6):331–46.

12. Lledo PM, Alonso M, Grubb MS. Adult neurogenesis and functional plasticity in neuronal circuits. Nat Rev Neurosci. 2006;7(3):179–93. doi:10.1038/nrn1867. nrn1867 [pii].
13. Petreanu L, Alvarez-Buylla A. Maturation and death of adult-born olfactory bulb granule neurons: role of olfaction. J Neurosci. 2002;22(14):6106–13.
14. Capilla-Gonzalez V, Gil-Perotin S, Ferragud A, Bonet-Ponce L, Canales JJ, Garcia-Verdugo JM. Exposure to N-ethyl-N-nitrosourea in adult mice alters structural and functional integrity of neurogenic sites. PLoS One. 2012;7(1):e29891. doi:10.1371/journal.pone.0029891. PONE-D-11-07990 [pii].
15. Sanai N, Berger MS, Garcia-Verdugo JM, Alvarez-Buylla A. Comment on "Human neuroblasts migrate to the olfactory bulb via a lateral ventricular extension". Science. 2007;318(5849):393. doi:10.1126/science.1145011. 318/5849/393b [pii].
16. Curtis MA, Kam M, Nannmark U, Anderson MF, Axell MZ, Wikkelso C, Holtas S, van Roon-Mom WM, Bjork-Eriksson T, Nordborg C, Frisen J, Dragunow M, Faull RL, Eriksson PS. Human neuroblasts migrate to the olfactory bulb via a lateral ventricular extension. Science. 2007;315(5816):1243–9. doi:10.1126/science.1136281. 1136281 [pii].
17. Wang C, Liu F, Liu YY, Zhao CH, You Y, Wang L, Zhang J, Wei B, Ma T, Zhang Q, Zhang Y, Chen R, Song H, Yang Z. Identification and characterization of neuroblasts in the subventricular zone and rostral migratory stream of the adult human brain. Cell Res. 2011;21(11):1534–50. doi:10.1038/cr.2011.83. cr201183 [pii].
18. Kam M, Curtis MA, McGlashan SR, Connor B, Nannmark U, Faull RL. The cellular composition and morphological organization of the rostral migratory stream in the adult human brain. J Chem Neuroanat. 2009;37(3):196–205. doi:10.1016/j.jchemneu.2008.12.009. S0891-0618(08)00165-8 [pii].
19. Sanai N, Tramontin AD, Quinones-Hinojosa A, Barbaro NM, Gupta N, Kunwar S, Lawton MT, McDermott MW, Parsa AT, Garcia-Verdugo JM, Berger MS, Alvarez-Buylla A. Unique astrocyte ribbon in adult human brain contains neural stem cells but lacks chain migration. Nature. 2004;427(6976):740–4. doi:10.1038/nature02301. nature02301 [pii].
20. Guerrero-Cazares H, Gonzalez-Perez O, Soriano-Navarro M, Zamora-Berridi G, Garcia-Verdugo JM, Quinones-Hinojosa A. Cytoarchitecture of the lateral ganglionic eminence and rostral extension of the lateral ventricle in the human fetal brain. J Comp Neurol. 2011;519(6):1165–80. doi:10.1002/cne.22566.
21. Quinones-Hinojosa A, Sanai N, Soriano-Navarro M, Gonzalez-Perez O, Mirzadeh Z, Gil-Perotin S, Romero-Rodriguez R, Berger MS, Garcia-Verdugo JM, Alvarez-Buylla A. Cellular composition and cytoarchitecture of the adult human subventricular zone: a niche of neural stem cells. J Comp Neurol. 2006;494(3):415–34. doi:10.1002/cne.20798.
22. Sanai N, Nguyen T, Ihrie RA, Mirzadeh Z, Tsai HH, Wong M, Gupta N, Berger MS, Huang E, Garcia-Verdugo JM, Rowitch DH, Alvarez-Buylla A. Corridors of migrating neurons in the human brain and their decline during infancy. Nature. 2011;478(7369):382–6. doi:10.1038/nature10487. nature10487 [pii].
23. Kaneko N, Marin O, Koike M, Hirota Y, Uchiyama Y, Wu JY, Lu Q, Tessier-Lavigne M, Alvarez-Buylla A, Okano H, Rubenstein JL, Sawamoto K. New neurons clear the path of astrocytic processes for their rapid migration in the adult brain. Neuron. 2010;67(2):213–23. doi:10.1016/j.neuron.2010.06.018.
24. Sawamoto K, Wichterle H, Gonzalez-Perez O, Cholfin JA, Yamada M, Spassky N, Murcia NS, Garcia-Verdugo JM, Marin O, Rubenstein JL, Tessier-Lavigne M, Okano H, Alvarez-Buylla A. New neurons follow the flow of cerebrospinal fluid in the adult brain. Science. 2006;311(5761):629–32.
25. Vukovic J, Blackmore DG, Jhaveri D, Bartlett PF. Activation of neural precursors in the adult neurogenic niches. Neurochem Int. 2011;59(3):341–6. doi:10.1016/j.neuint.2011.04.003. S0197-0186(11)00156-2 [pii].
26. Christie KJ, Turnley AM. Regulation of endogenous neural stem/progenitor cells for neural repair-factors that promote neurogenesis and gliogenesis in the normal and damaged brain. Front Cell Neurosci. 2012;6:70. doi:10.3389/fncel.2012.00070.

27. Nait-Oumesmar B, Decker L, Lachapelle F, Avellana-Adalid V, Bachelin C, Van Evercooren AB. Progenitor cells of the adult mouse subventricular zone proliferate, migrate and differentiate into oligodendrocytes after demyelination. Eur J Neurosci. 1999;11(12):4357–66. doi:ejn873 [pii].
28. Marti-Fabregas J, Romaguera-Ros M, Gomez-Pinedo U, Martinez-Ramirez S, Jimenez-Xarrie E, Marin R, Marti-Vilalta JL, Garcia-Verdugo JM. Proliferation in the human ipsilateral subventricular zone after ischemic stroke. Neurology. 2010;74(5):357–65. doi:10.1212/WNL.0b013e3181cbccec. WNL.0b013e3181cbccec [pii].
29. Yamashita T, Ninomiya M, Hernandez Acosta P, Garcia-Verdugo JM, Sunabori T, Sakaguchi M, Adachi K, Kojima T, Hirota Y, Kawase T, Araki N, Abe K, Okano H, Sawamoto K. Subventricular zone-derived neuroblasts migrate and differentiate into mature neurons in the post-stroke adult striatum. J Neurosci. 2006;26(24):6627–36.
30. Barkho BZ, Zhao X. Adult neural stem cells: response to stroke injury and potential for therapeutic applications. Curr Stem Cell Res Ther. 2011;6(4):327–38. doi:BSP/CSCRT/E-Pub/00090 [pii].
31. Capilla-Gonzalez V, Guerrero-Cazares H, Bonsu JM, Gonzalez-Perez O, Achanta P, Wong J, Garcia-Verdugo JM, Quinones-Hinojosa A. The subventricular zone is able to respond to a demyelinating lesion after localized radiation. Stem Cells. 2014;32(1):59–69. doi:10.1002/stem.1519.
32. Doetsch F, Caille I, Lim DA, Garcia-Verdugo JM, Alvarez-Buylla A. Subventricular zone astrocytes are neural stem cells in the adult mammalian brain. Cell. 1999;97(6):703–16.
33. Ihrie RA, Alvarez-Buylla A. Cells in the astroglial lineage are neural stem cells. Cell Tissue Res. 2008;331(1):179–91. doi:10.1007/s00441-007-0461-z.
34. Morrens J, Van Den Broeck W, Kempermann G. Glial cells in adult neurogenesis. Glia. 2012;60(2):159–74. doi:10.1002/glia.21247.
35. Doetsch F, Garcia-Verdugo JM, Alvarez-Buylla A. Cellular composition and three-dimensional organization of the subventricular germinal zone in the adult mammalian brain. J Neurosci. 1997;17(13):5046–61.
36. Han YG, Spassky N, Romaguera-Ros M, Garcia-Verdugo JM, Aguilar A, Schneider-Maunoury S, Alvarez-Buylla A. Hedgehog signaling and primary cilia are required for the formation of adult neural stem cells. Nat Neurosci. 2008;11(3):277–84. doi:10.1038/nn2059. nn2059 [pii].
37. Mirzadeh Z, Merkle FT, Soriano-Navarro M, Garcia-Verdugo JM, Alvarez-Buylla A. Neural stem cells confer unique pinwheel architecture to the ventricular surface in neurogenic regions of the adult brain. Cell Stem Cell. 2008;3(3):265–78.
38. Gil-Perotin S, Alvarez-Buylla A, Garcia-Verdugo JM. Identification and characterization of neural progenitor cells in the adult mammalian brain. Adv Anat Embryol Cell Biol. 2009;203:1–101. ix.
39. Ming GL, Song HJ. Adult neurogenesis in the mammalian central nervous system. Annu Rev Neurosci. 2005;28:223–50. doi:10.1146/annurev.neuro.28.051804.101459.
40. Gleeson JG, Lin PT, Flanagan LA, Walsh CA. Doublecortin is a microtubule-associated protein and is expressed widely by migrating neurons. Neuron. 1999;23(2):257–71.
41. Didier M, Harandi M, Aguera M, Bancel B, Tardy M, Fages C, Calas A, Stagaard M, Mollgard K, Belin MF. Differential immunocytochemical staining for glial fibrillary acidic (GFA) protein, S-100 protein and glutamine synthetase in the rat subcommissural organ, nonspecialized ventricular ependyma and adjacent neuropil. Cell Tissue Res. 1986;245(2):343–51.
42. Craig CG, Tropepe V, Morshead CM, Reynolds BA, Weiss S, van der Kooy D. In vivo growth factor expansion of endogenous subependymal neural precursor cell populations in the adult mouse brain. J Neurosci. 1996;16(8):2649–58.
43. Spassky N, Merkle FT, Flames N, Tramontin AD, Garcia-Verdugo JM, Alvarez-Buylla A. Adult ependymal cells are postmitotic and are derived from radial glial cells during embryogenesis. J Neurosci. 2005;25(1):10–8.

44. Alvarez-Buylla A, Garcia-Verdugo JM. Neurogenesis in adult subventricular zone. J Neurosci. 2002;22(3):629–34.
45. Peretto P, Merighi A, Fasolo A, Bonfanti L. Glial tubes in the rostral migratory stream of the adult rat. Brain Res Bull. 1997;42(1):9–21. doi:S0361-9230(96)00116-5 [pii].
46. Pencea V, Bingaman KD, Freedman LJ, Luskin MB. Neurogenesis in the subventricular zone and rostral migratory stream of the neonatal and adult primate forebrain. Exp Neurol. 2001;172(1):1–16. doi:10.1006/exnr.2001.7768.
47. De Marchis S, Fasolo A, Puche AC. Subventricular zone-derived neuronal progenitors migrate into the subcortical forebrain of postnatal mice. J Comp Neurol. 2004;476(3): 290–300. doi:10.1002/cne.20217.
48. Sawada M, Sawamoto K. Mechanisms of neurogenesis in the normal and injured adult brain. Keio J Med. 2013;62(1):13–28.
49. Porlan E, Perez-Villalba A, Delgado AC, Ferron SR. Paracrine regulation of neural stem cells in the subependymal zone. Arch Biochem Biophys. 2013;534(1–2):11–9. doi:10.1016/j.abb.2012.10.001. S0003-9861(12)00360-8 [pii].
50. Pencea V, Luskin MB. Prenatal development of the rodent rostral migratory stream. J Comp Neurol. 2003;463(4):402–18. doi:10.1002/cne.10746.
51. Mobley AS, Bryant AK, Richard MB, Brann JH, Firestein SJ, Greer CA. Age-dependent regional changes in the rostral migratory stream. Neurobiol Aging. 2013;34(7):1873–81. doi:10.1016/j.neurobiolaging.2013.01.015. S0197-4580(13)00045-6 [pii].
52. Namihira M, Nakashima K. Mechanisms of astrocytogenesis in the mammalian brain. Curr Opin Neurobiol. 2013;23(6):921–7. doi:10.1016/j.conb.2013.06.002. S0959-4388(13)00121-9 [pii].
53. Benner EJ, Luciano D, Jo R, Abdi K, Paez-Gonzalez P, Sheng H, Warner DS, Liu C, Eroglu C, Kuo CT. Protective astrogenesis from the SVZ niche after injury is controlled by Notch modulator Thbs4. Nature. 2013;497(7449):369–73. doi:10.1038/nature12069. nature12069 [pii].
54. Mejia-Gervacio S, Murray K, Lledo PM. NKCC1 controls GABAergic signaling and neuroblast migration in the postnatal forebrain. Neural Dev. 2011;6:4. doi:10.1186/1749-8104-6-4. 1749-8104-6-4 [pii].
55. Rosengren S, Henson PM, Worthen GS. Migration-associated volume changes in neutrophils facilitate the migratory process in vitro. Am J Physiol. 1994;267(6 Pt 1):C1623–32.
56. Rotte A, Pasham V, Yang W, Eichenmuller M, Bhandaru M, Shumilina E, Lang F. Phosphoinositide 3-kinase-dependent regulation of Na+/H+ exchanger in dendritic cells. Pflugers Arch. 2010;460(6):1087–96. doi:10.1007/s00424-010-0879-0.
57. Klein M, Seeger P, Schuricht B, Alper SL, Schwab A. Polarization of Na(+)/H(+) and Cl(−)/HCO (3)(−) exchangers in migrating renal epithelial cells. J Gen Physiol. 2000;115(5): 599–608.
58. Horesh D, Sapir T, Francis F, Wolf SG, Caspi M, Elbaum M, Chelly J, Reiner O. Doublecortin, a stabilizer of microtubules. Hum Mol Genet. 1999;8(9):1599–610.
59. Gleeson JG, Allen KM, Fox JW, Lamperti ED, Berkovic S, Scheffer I, Cooper EC, Dobyns WB, Minnerath SR, Ross ME, Walsh CA. doublecortin, a brain-specific gene mutated in human X-linked lissencephaly and double cortex syndrome, encodes a putative signaling protein. Cell. 1998;92(1):63–72. doi:10.1016/s0092-8674(00)80899-5.
60. Francis F, Koulakoff A, Boucher D, Chafey P, Schaar B, Vinet MC, Friocourt G, McDonnell N, Reiner O, Kahn A, McConnell SK, Berwald-Netter Y, Denoulet P, Chelly J. Doublecortin is a developmentally regulated, microtubule-associated protein expressed in migrating and differentiating neurons. Neuron. 1999;23(2):247–56. doi:S0896-6273(00)80777-1 [pii].
61. Ocbina PJ, Dizon ML, Shin L, Szele FG. Doublecortin is necessary for the migration of adult subventricular zone cells from neurospheres. Mol Cell Neurosci. 2006;33(2):126–35. doi:10.1016/j.mcn.2006.06.014. S1044-7431(06)00134-5 [pii].
62. Koizumi H, Higginbotham H, Poon T, Tanaka T, Brinkman BC, Gleeson JG. Doublecortin maintains bipolar shape and nuclear translocation during migration in the adult forebrain. Nat Neurosci. 2006;9(6):779–86. doi:10.1038/nn1704. nn1704 [pii].

63. Kappeler C, Saillour Y, Baudoin JP, Tuy FP, Alvarez C, Houbron C, Gaspar P, Hamard G, Chelly J, Metin C, Francis F. Branching and nucleokinesis defects in migrating interneurons derived from doublecortin knockout mice. Hum Mol Genet. 2006;15(9):1387–400.
64. Sontheimer H. An unexpected role for ion channels in brain tumor metastasis. Exp Biol Med (Maywood). 2008;233(7):779–91. doi:10.3181/0711-MR-308. 0711-MR-308 [pii].
65. Gamba G. Molecular physiology and pathophysiology of electroneutral cation-chloride cotransporters. Physiol Rev. 2005;85(2):423–93. doi:10.1152/physrev.00011.2004. 85/2/423 [pii].
66. Lytle C, Forbush 3rd B. Regulatory phosphorylation of the secretory Na-K-Cl cotransporter: modulation by cytoplasmic Cl. Am J Physiol. 1996;270(2 Pt 1):C437–48.
67. Yamada J, Okabe A, Toyoda H, Kilb W, Luhmann HJ, Fukuda A. Cl-uptake promoting depolarizing GABA actions in immature rat neocortical neurones is mediated by NKCC1. J Physiol. 2004;557(Pt 3):829–41. doi:10.1113/jphysiol.2004.062471. jphysiol.2004.062471 [pii].
68. Russell JM. Sodium-potassium-chloride cotransport. Physiol Rev. 2000;80(1):211–76.
69. Strange K. Cellular volume homeostasis. Adv Physiol Educ. 2004;28(1–4):155–9. doi:10.1152/advan.00034.2004. 28/4/155 [pii].
70. Garzon-Muvdi T, Schiapparelli P, Rhys C, Guerrero-Cazares H, Smith C, Kim DH, Kone L, Farber H, Lee DY, An SS, Levchenko A, Quinones-Hinojosa A. Regulation of brain tumor dispersal by NKCC1 through a novel role in focal adhesion regulation. PLoS Biol. 2012;10(5):e1001320. doi:10.1371/journal.pbio.1001320. PBIOLOGY-D-11-02751 [pii].
71. Ohshima T, Ward JM, Huh CG, Longenecker G, Veeranna PHC, Brady RO, Martin LJ, Kulkarni AB. Targeted disruption of the cyclin-dependent kinase 5 gene results in abnormal corticogenesis, neuronal pathology and perinatal death. Proc Natl Acad Sci U S A. 1996; 93(20):11173–8.
72. Gilmore EC, Ohshima T, Goffinet AM, Kulkarni AB, Herrup K. Cyclin-dependent kinase 5-deficient mice demonstrate novel developmental arrest in cerebral cortex. J Neurosci. 1998;18(16):6370–7.
73. Ohshima T, Hirasawa M, Tabata H, Mutoh T, Adachi T, Suzuki H, Saruta K, Iwasato T, Itohara S, Hashimoto M, Nakajima K, Ogawa M, Kulkarni AB, Mikoshiba K. Cdk5 is required for multipolar-to-bipolar transition during radial neuronal migration and proper dendrite development of pyramidal neurons in the cerebral cortex. Development. 2007;134(12):2273–82. doi:10.1242/dev.02854. dev.02854 [pii].
74. Hirota Y, Ohshima T, Kaneko N, Ikeda M, Iwasato T, Kulkarni AB, Mikoshiba K, Okano H, Sawamoto K. Cyclin-dependent kinase 5 is required for control of neuroblast migration in the postnatal subventricular zone. J Neurosci. 2007;27(47):12829–38. doi:10.1523/JNEUROSCI.1014-07.2007. 27/47/12829 [pii].
75. Bonfanti L, Olive S, Poulain DA, Theodosis DT. Mapping of the distribution of polysialylated neural cell adhesion molecule throughout the central nervous system of the adult rat: an immunohistochemical study. Neuroscience. 1992;49(2):419–36. doi:0306-4522(92)90107-D [pii].
76. Bonfanti L. PSA-NCAM in mammalian structural plasticity and neurogenesis. Prog Neurobiol. 2006;80(3):129–64. doi:10.1016/j.pneurobio.2006.08.003.
77. Yang P, Yin X, Rutishauser U. Intercellular space is affected by the polysialic acid content of NCAM. J Cell Biol. 1992;116(6):1487–96.
78. Hu H, Tomasiewicz H, Magnuson T, Rutishauser U. The role of polysialic acid in migration of olfactory bulb interneuron precursors in the subventricular zone. Neuron. 1996;16(4):735–43. doi:S0896-6273(00)80094-X [pii].
79. Bonfanti L, Theodosis DT. Expression of polysialylated neural cell adhesion molecule by proliferating cells in the subependymal layer of the adult rat, in its rostral extension and in the olfactory bulb. Neuroscience. 1994;62(1):291–305. doi:0306-4522(94)90333-6 [pii].
80. Cremer H, Lange R, Christoph A, Plomann M, Vopper G, Roes J, Brown R, Baldwin S, Kraemer P, Scheff S, et al. Inactivation of the N-CAM gene in mice results in size reduction

of the olfactory bulb and deficits in spatial learning. Nature. 1994;367(6462):455–9. doi:10.1038/367455a0.
81. Chazal G, Durbec P, Jankovski A, Rougon G, Cremer H. Consequences of neural cell adhesion molecule deficiency on cell migration in the rostral migratory stream of the mouse. J Neurosci. 2000;20(4):1446–57.
82. Ono K, Tomasiewicz H, Magnuson T, Rutishauser U. N-CAM mutation inhibits tangential neuronal migration and is phenocopied by enzymatic removal of polysialic acid. Neuron. 1994;13(3):595–609. doi:0896-6273(94)90028-0 [pii].
83. Plow EF, Haas TA, Zhang L, Loftus J, Smith JW. Ligand binding to integrins. J Biol Chem. 2000;275(29):21785–8. doi:10.1074/jbc.R000003200. R000003200 [pii].
84. Murase S, Horwitz AF. Deleted in colorectal carcinoma and differentially expressed integrins mediate the directional migration of neural precursors in the rostral migratory stream. J Neurosci. 2002;22(9):3568–79. doi:20026349. 22/9/3568 [pii].
85. Emsley JG, Hagg T. alpha6beta1 integrin directs migration of neuronal precursors in adult mouse forebrain. Exp Neurol. 2003;183(2):273–85. doi:S0014488603002097 [pii].
86. Mobley AK, McCarty JH. beta8 integrin is essential for neuroblast migration in the rostral migratory stream. Glia. 2011;59(11):1579–87. doi:10.1002/glia.21199.
87. Tessier-Lavigne M. Wiring the brain: the logic and molecular mechanisms of axon guidance and regeneration. Harvey Lect. 2002;98:103–43.
88. Staquicini FI, Dias-Neto E, Li J, Snyder EY, Sidman RL, Pasqualini R, Arap W. Discovery of a functional protein complex of netrin-4, laminin gamma1 chain, and integrin alpha6beta1 in mouse neural stem cells. Proc Natl Acad Sci U S A. 2009;106(8):2903–8. doi:10.1073/pnas.0813286106. 0813286106 [pii].
89. Ingham PW, McMahon AP. Hedgehog signaling in animal development: paradigms and principles. Genes Dev. 2001;15(23):3059–87. doi:10.1101/gad.938601.
90. Angot E, Loulier K, Nguyen-Ba-Charvet KT, Gadeau AP, Ruat M, Traiffort E. Chemoattractive activity of sonic hedgehog in the adult subventricular zone modulates the number of neural precursors reaching the olfactory bulb. Stem Cells. 2008;26(9):2311–20. doi:10.1634/stemcells. 2008-0297. 2008-0297 [pii].
91. Hor CH, Tang BL. Sonic hedgehog as a chemoattractant for adult NPCs. Cell Adh Migr. 2010;4(1):1–3. doi:9914 [pii].
92. Ihrie RA, Shah JK, Harwell CC, Levine JH, Guinto CD, Lezameta M, Kriegstein AR, Alvarez-Buylla A. Persistent sonic hedgehog signaling in adult brain determines neural stem cell positional identity. Neuron. 2011;71(2):250–62. doi:10.1016/j.neuron.2011.05.018. S0896-6273(11)00404-1 [pii].
93. Nguyen-Ba-Charvet KT, Chedotal A. Role of Slit proteins in the vertebrate brain. J Physiol Paris. 2002;96(1–2):91–8. doi:10.1016/s0928-4257(01)00084-5. Pii s0928-4257(01)00084-5.
94. Wong K, Park HT, Wu JY, Rao Y. Slit proteins: molecular guidance cues for cells ranging from neurons to leukocytes. Curr Opin Genet Dev. 2002;12(5):583–91. doi:10.1016/s0959-437x(02)00343-x.
95. Chilton JK. Molecular mechanisms of axon guidance. Dev Biol. 2006;292(1):13–24. doi:10.1016/j.ydbio.2005.12.048.
96. Nguyen-Ba-Charvet KT, Picard-Riera N, Tessier-Lavigne M, Baron-Van Evercooren A, Sotelo C, Chedotal A. Multiple roles for slits in the control of cell migration in the rostral migratory stream. J Neurosci. 2004;24(6):1497–506. doi:10.1523/jneurosci. 4729-03.2004.
97. Kitabgi P, Melik-Parsadaniantz S, Rostene W (2006) Chemokines: a new peptide family of neuromodulators. Handbook of Biologically Active Peptides. p. 559–565. doi:10.1016/B978-012369442-3/50083-0.
98. Schonemeier B, Kolodziej A, Schulz S, Jacobs S, Hoellt V, Stumm R. Regional and cellular localization of the CXCl12/SDF-1 chemokine receptor CXCR7 in the developing and adult rat brain. J Comp Neurol. 2008;510(2):207–20. doi:10.1002/cne.21780.

99. Odemis V, Boosmann K, Heinen A, Kury P, Engele J. CXCR7 is an active component of SDF-1 signalling in astrocytes and Schwann cells. J Cell Sci. 2010;123(Pt 7):1081–8. doi:10.1242/jcs.062810.
100. Tiveron MC, Boutin C, Daou P, Moepps B, Cremer H. Expression and function of CXCR7 in the mouse forebrain. J Neuroimmunol. 2010;224(1–2):72–9.
101. Kojima T, Hirota Y, Ema M, Takahashi S, Miyoshi I, Okano H, Sawamoto K. Subventricular zone-derived neural progenitor cells migrate along a blood vessel scaffold toward the post-stroke striatum. Stem Cells. 2010;28(3):545–54. doi:10.1002/stem.306.
102. Arvidsson A, Collin T, Kirik D, Kokaia Z, Lindvall O. Neuronal replacement from endogenous precursors in the adult brain after stroke. Nat Med. 2002;8(9):963–70. doi:10.1038/nm747. nm747 [pii].
103. Liu BF, Gao EJ, Zeng XZ, Ji M, Cai Q, Lu Q, Yang H, Xu QY. Proliferation of neural precursors in the subventricular zone after chemical lesions of the nigrostriatal pathway in rat brain. Brain Res. 2006;1106(1):30–9. doi:10.1016/j.brainres.2006.05.111. S0006-8993(06)01534-4 [pii].
104. Nait-Oumesmar B, Picard-Riera N, Kerninon C, Decker L, Seilhean D, Hoglinger GU, Hirsch EC, Reynolds R, Baron-Van Evercooren A. Activation of the subventricular zone in multiple sclerosis: evidence for early glial progenitors. Proc Natl Acad Sci U S A. 2007;104(11):4694–9.
105. Otero L, Zurita M, Bonilla C, Rico MA, Aguayo C, Rodriguez A, Vaquero J. Endogenous neurogenesis after intracerebral hemorrhage. Histol Histopathol. 2012;27(3):303–15.
106. Del Carmen Gomez-Roldan M, Perez-Martin M, Capilla-Gonzalez V, Cifuentes M, Perez J, Garcia-Verdugo JM, Fernandez-Llebrez P. Neuroblast proliferation on the surface of the adult rat striatal wall after focal ependymal loss by intracerebroventricular injection of neuraminidase. J Comp Neurol. 2008;507(4):1571–87. doi:10.1002/cne.21618.
107. Menn B, Garcia-Verdugo JM, Yaschine C, Gonzalez-Perez O, Rowitch D, Alvarez-Buylla A. Origin of oligodendrocytes in the subventricular zone of the adult brain. J Neurosci. 2006;26(30):7907–18. doi:10.1523/JNEUROSCI. 1299-06.2006. 26/30/7907 [pii].
108. Picard-Riera N, Decker L, Delarasse C, Goude K, Nait-Oumesmar B, Liblau R, Pham-Dinh D, Evercooren AB. Experimental autoimmune encephalomyelitis mobilizes neural progenitors from the subventricular zone to undergo oligodendrogenesis in adult mice. Proc Natl Acad Sci U S A. 2002;99(20):13211–6. doi:10.1073/pnas.192314199. 192314199 [pii].
109. Gonzalez-Perez O, Romero-Rodriguez R, Soriano-Navarro M, Garcia-Verdugo JM, Alvarez-Buylla A. Epidermal growth factor induces the progeny of subventricular zone type B cells to migrate and differentiate into oligodendrocytes. Stem Cells. 2009;27(8):2032–43. doi:10.1002/stem.119.
110. Kaneko N, Sawamoto K. Neuronal migration in the adult brain. Nihon Shinkei Seishin Yakurigaku Zasshi. 2007;27(5–6):215–8.
111. Gonzalez-Perez O, Gutierrez-Fernandez F, Lopez-Virgen V, Collas-Aguilar J, Quinones-Hinojosa A, Garcia-Verdugo JM. Immunological regulation of neurogenic niches in the adult brain. Neuroscience. 2012;226:270–81. doi:10.1016/j.neuroscience.2012.08.053. S0306-4522(12)00888-3 [pii].
112. Logan TT, Villapol S, Symes AJ. TGF-beta superfamily gene expression and induction of the Runx1 transcription factor in adult neurogenic regions after brain injury. PLoS One. 2013;8(3):e59250. doi:10.1371/journal.pone.0059250. PONE-D-12-27158 [pii].
113. Yan YP, Sailor KA, Vemuganti R, Dempsey RJ. Insulin-like growth factor-1 is an endogenous mediator of focal ischemia-induced neural progenitor proliferation. Eur J Neurosci. 2006;24(1):45–54. doi:10.1111/j.1460-9568.2006.04872.x. EJN4872 [pii].
114. Kang SS, Keasey MP, Arnold SA, Reid R, Geralds J, Hagg T. Endogenous CNTF mediates stroke-induced adult CNS neurogenesis in mice. Neurobiol Dis. 2012;49C:68–78. doi:10.1016/j.nbd.2012.08.020. S0969-9961(12)00307-5 [pii].
115. Pluchino S, Muzio L, Imitola J, Deleidi M, Alfaro-Cervello C, Salani G, Porcheri C, Brambilla E, Cavasinni F, Bergamaschi A, Garcia-Verdugo JM, Comi G, Khoury SJ,

Martino G. Persistent inflammation alters the function of the endogenous brain stem cell compartment. Brain. 2008;131(Pt 10):2564–78. doi:10.1093/brain/awn198. awn198 [pii].
116. Imitola J, Raddassi K, Park KI, Mueller FJ, Nieto M, Teng YD, Frenkel D, Li J, Sidman RL, Walsh CA, Snyder EY, Khoury SJ. Directed migration of neural stem cells to sites of CNS injury by the stromal cell-derived factor 1alpha/CXC chemokine receptor 4 pathway. Proc Natl Acad Sci U S A. 2004;101(52):18117–22. doi:10.1073/pnas.0408258102. 0408258102 [pii].
117. Tran PB, Ren D, Veldhouse TJ, Miller RJ. Chemokine receptors are expressed widely by embryonic and adult neural progenitor cells. J Neurosci Res. 2004;76(1):20–34. doi:10.1002/jnr.20001.
118. Jaerve A, Muller HW. Chemokines in CNS injury and repair. Cell Tissue Res. 2012; 349(1):229–48. doi:10.1007/s00441-012-1427-3.
119. Turbic A, Leong SY, Turnley AM. Chemokines and inflammatory mediators interact to regulate adult murine neural precursor cell proliferation, survival and differentiation. PLoS One. 2011;6(9):e25406. doi:10.1371/journal.pone.0025406. PONE-D-11-06373 [pii].
120. Belmadani A, Tran PB, Ren D, Miller RJ. Chemokines regulate the migration of neural progenitors to sites of neuroinflammation. J Neurosci. 2006;26(12):3182–91. doi:10.1523/JNEUROSCI. 0156-06.2006. 26/12/3182 [pii].
121. Gordon RJ, Mehrabi NF, Maucksch C, Connor B. Chemokines influence the migration and fate of neural precursor cells from the young adult and middle-aged rat subventricular zone. Experimental neurology. 2012;233(1):587–94. doi:10.1016/j.expneurol.2011.11.029. S0014-4886(11)00433-X [pii].
122. Widera D, Holtkamp W, Entschladen F, Niggemann B, Zanker K, Kaltschmidt B, Kaltschmidt C. MCP-1 induces migration of adult neural stem cells. Eur J Cell Biol. 2004;83(8):381–7. doi:10.1078/0171-9335-00403. S0171-9335(04)70406-9 [pii].
123. Gordon RJ, McGregor AL, Connor B. Chemokines direct neural progenitor cell migration following striatal cell loss. Mol Cell Neurosci. 2009;41(2):219–32. doi:10.1016/j.mcn.2009.03.001. S1044-7431(09)00061-X [pii].
124. Liu XS, Zhang ZG, Zhang RL, Gregg SR, Wang L, Yier T, Chopp M. Chemokine ligand 2 (CCL2) induces migration and differentiation of subventricular zone cells after stroke. J Neurosci Res. 2007;85(10):2120–5. doi:10.1002/jnr.21359.
125. Yan YP, Sailor KA, Lang BT, Park SW, Vemuganti R, Dempsey RJ. Monocyte chemoattractant protein-1 plays a critical role in neuroblast migration after focal cerebral ischemia. J Cereb Blood Flow Metab. 2007;27(6):1213–24. doi:10.1038/sj.jcbfm.9600432. 9600432 [pii].
126. Saha B, Peron S, Murray K, Jaber M, Gaillard A. Cortical lesion stimulates adult subventricular zone neural progenitor cell proliferation and migration to the site of injury. Stem Cell Res. 2013;11(3):965–77. doi:10.1016/j.scr.2013.06.006. S1873-5061(13)00079-2 [pii].
127. Thored P, Arvidsson A, Cacci E, Ahlenius H, Kallur T, Darsalia V, Ekdahl CT, Kokaia Z, Lindvall O. Persistent production of neurons from adult brain stem cells during recovery after stroke. Stem Cells. 2006;24(3):739–47. doi:10.1634/stemcells. 2005-0281. 2005–0281 [pii].
128. Robin AM, Zhang ZG, Wang L, Zhang RL, Katakowski M, Zhang L, Wang Y, Zhang C, Chopp M. Stromal cell-derived factor 1alpha mediates neural progenitor cell motility after focal cerebral ischemia. J Cereb Blood Flow Metab. 2006;26(1):125–34. doi:10.1038/sj.jcbfm.9600172. 9600172 [pii].
129. Itoh T, Satou T, Ishida H, Nishida S, Tsubaki M, Hashimoto S, Ito H. The relationship between SDF-1alpha/CXCR4 and neural stem cells appearing in damaged area after traumatic brain injury in rats. Neurol Res. 2009;31(1):90–102. doi:10.1179/174313208X332995.
130. Aguirre A, Gallo V. Reduced EGFR signaling in progenitor cells of the adult subventricular zone attenuates oligodendrogenesis after demyelination. Neuron Glia Biol. 2007;3(3): 209–20. doi:10.1017/S1740925X08000082. S1740925X08000082 [pii].
131. Decker L, Picard-Riera N, Lachapelle F, Baron-Van Evercooren A. Growth factor treatment promotes mobilization of young but not aged adult subventricular zone precursors in response to demyelination. J Neurosci Res. 2002;69(6):763–71. doi:10.1002/jnr.10411.

132. Lachapelle F, Avellana-Adalid V, Nait-Oumesmar B, Baron-Van Evercooren A. Fibroblast growth factor-2 (FGF-2) and platelet-derived growth factor AB (PDGF AB) promote adult SVZ-derived oligodendrogenesis in vivo. Mol Cell Neurosci. 2002;20(3):390–403. doi:S1044743102911243 [pii].
133. Wang Y, Jin K, Mao XO, Xie L, Banwait S, Marti HH, Greenberg DA. VEGF-overexpressing transgenic mice show enhanced post-ischemic neurogenesis and neuromigration. J Neurosci Res. 2007;85(4):740–7. doi:10.1002/jnr.21169.
134. Bao X, Wei J, Feng M, Lu S, Li G, Dou W, Ma W, Ma S, An Y, Qin C, Zhao RC, Wang R. Transplantation of human bone marrow-derived mesenchymal stem cells promotes behavioral recovery and endogenous neurogenesis after cerebral ischemia in rats. Brain Res. 2011;1367:103–13. doi:10.1016/j.brainres.2010.10.063. S0006-8993(10)02332-2 [pii].
135. Harms KM, Li L, Cunningham LA. Murine neural stem/progenitor cells protect neurons against ischemia by HIF-1alpha-regulated VEGF signaling. PLoS One. 2010;5(3):e9767. doi:10.1371/journal.pone.0009767.
136. Barkho BZ, Munoz AE, Li X, Li L, Cunningham LA, Zhao X. Endogenous matrix metalloproteinase (MMP)-3 and MMP-9 promote the differentiation and migration of adult neural progenitor cells in response to chemokines. Stem Cells. 2008;26(12):3139–49. doi:10.1634/stemcells. 2008-0519. 2008-0519 [pii].
137. Courtes S, Vernerey J, Pujadas L, Magalon K, Cremer H, Soriano E, Durbec P, Cayre M. Reelin controls progenitor cell migration in the healthy and pathological adult mouse brain. PLoS One. 2011;6(5):e20430. doi:10.1371/journal.pone.0020430. PONE-D-11-03174 [pii].
138. Aboody KS, Najbauer J, Danks MK. Stem and progenitor cell-mediated tumor selective gene therapy. Gene Ther. 2008;15(10):739–52. doi:10.1038/gt.2008.41.
139. McGirt MJ, Chaichana KL, Gathinji M, Attenello FJ, Than K, Olivi A, Weingart JD, Brem H, Quinones-Hinojosa AR. Independent association of extent of resection with survival in patients with malignant brain astrocytoma. J Neurosurg. 2009;110(1):156–62. doi:10.3171/2008.4.17536.
140. Stupp R, Mason WP, van den Bent MJ, Weller M, Fisher B, Taphoorn MJ, Belanger K, Brandes AA, Marosi C, Bogdahn U, Curschmann J, Janzer RC, Ludwin SK, Gorlia T, Allgeier A, Lacombe D, Cairncross JG, Eisenhauer E, Mirimanoff RO. Radiotherapy plus concomitant and adjuvant temozolomide for glioblastoma. N Engl J Med. 2005;352(10):987–96. doi:10.1056/NEJMoa043330.
141. Aboody KS, Brown A, Rainov NG, Bower KA, Liu S, Yang W, Small JE, Herrlinger U, Ourednik V, Black PM, Breakefield XO, Snyder EY. Neural stem cells display extensive tropism for pathology in adult brain: evidence from intracranial gliomas. Proc Natl Acad Sci U S A. 2000;97(23):12846–51. doi:10.1073/pnas.97.23.12846.
142. Zhao D, Najbauer J, Garcia E, Metz MZ, Gutova M, Glackin CA, Kim SU, Aboody KS. Neural stem cell tropism to glioma: critical role of tumor hypoxia. Mol Cancer Res. 2008;6(12):1819–29. doi:10.1158/1541-7786.MCR-08-0146.
143. Ziu M, Schmidt NO, Cargioli TG, Aboody KS, Black PM, Carroll RS. Glioma-produced extracellular matrix influences brain tumor tropism of human neural stem cells. J Neurooncol. 2006;79(2):125–33. doi:10.1007/s11060-006-9121-5.
144. Kendall SE, Najbauer J, Johnston HF, Metz MZ, Li S, Bowers M, Garcia E, Kim SU, Barish ME, Aboody KS, Glackin CA. Neural stem cell targeting of glioma is dependent on phosphoinositide 3-kinase signaling. Stem Cells. 2008;26(6):1575–86. doi:10.1634/stemcells. 2007-0887.
145. Frank RT, Najbauer J, Aboody KS. Concise review: stem cells as an emerging platform for antibody therapy of cancer. Stem Cells. 2010;28(11):2084–7. doi:10.1002/stem.513.
146. Aboody KS, Najbauer J, Metz MZ, D'Apuzzo M, Gutova M, Annala AJ, Synold TW, Couture LA, Blanchard S, Moats RA, Garcia E, Aramburo S, Valenzuela VV, Frank RT, Barish ME, Brown CE, Kim SU, Badie B, Portnow J. Neural stem cell-mediated enzyme/prodrug therapy for glioma: preclinical studies. Sci Transl Med. 2013;5(184):184ra159. doi:10.1126/scitranslmed.3005365.

The SVZ and Its Relationship to Stem Cell Based Neuro-oncogenesis

Yael Kusne and Nader Sanai

Abstract Gliomas are primary cancers of the brain and the most lethal cancers known to man. In recent years the discovery of germinal regions in the postnatal brain containing neuronal stem and progenitor cell populations has led to the hypothesis that these cells may themselves serve as an origin of brain tumors. Stem cells that reside within the glioma tumor have been shown to display nonneoplastic stem-like characteristics, including expression of various stem cell markers, as well as capacity for self-renewal and multipotency. Furthermore, glioma tumors display marked similarities to the germinal regions of the brain. Investigations of human neural stem cells and their potential for malignancy may finally identify a cell-of-origin for human gliomas. This, in turn, may facilitate better therapeutic targeting leading to improved prognosis for glioma patients.

Keywords Subventricular zone (SVZ) • Neuro-oncogenesis • Glioma cell of origin • Neural stem cells and glioma • Germinal regions and oncogenesis • Neurogenesis and oncogenesis of glioma

Introduction

Despite progress in research on the molecular aspects of malignant gliomas, the prognosis of these primary brain tumors continues to be dismal. In grade IV glioma, or glioblastoma, the most common glioma in adults, the median survival has changed only slightly in the last decade, increasing from 9 to 12 months in 2005 to the current median survival of 13–14 months. One reason for the lack of clinical advances is ignorance of the cellular origin of this disease, which delays the application of molecular analyses to treatment and impairs the ability to anticipate tumor behavior reliably.

Y. Kusne • N. Sanai (✉)
Barrow Brain Tumor Research Center, 350 W. Thomas Road, Phoenix, AZ 85013, USA
e-mail: Nader.Sanai@bnaneuro.net

M. Ehtesham (ed.), *Stem Cell Biology in Neoplasms of the Central Nervous System*, Advances in Experimental Medicine and Biology 853,
DOI 10.1007/978-3-319-16537-0_2

Historically, the neoplastic transformation of fully differentiated glia was widely assumed to be the only mechanism for gliomagenesis. Astrocytes and oligodendrocytes, once thought to be the sole dividing cells in the postnatal brain, were assumed to represent the cellular component most susceptible to transformation. More recently, however, this hypothesis has been challenged by the discovery of stem cell and progenitor populations residing in the postnatal brain, which may themselves serve as an origin of brain tumors. Phenotypic and behavioral similarities between gliomas and adult neural stem cells raise the possibility that stem or progenitor cells can give rise to gliomas. Resident tumor glioma stem cells display adult neural stem cell characteristics, including expression of stem cell markers (e.g., nestin), the ability to self-renew, and conserved multipotent potential. Candidate cells-of-origin include astrocytic neural stem cells (B cells) or transient amplifying precursors (C cells) of the adult subventricular zone (SVZ) and glial progenitor cells of the subcortical white matter. While a direct link remains to be established between any one of these cellular compartments and the formation of gliomas, recent advances have provided ample evidence to support the hypothesis.

Shared Features of Adult Germinal Regions and Gliomas

Adult germinal regions, such as the SVZ, are restricted to specialized microenvironments which allow for the survival and regulation of neural stem cells. Such specialized microenvironments consist of structural and molecular elements resulting in the appropriate conditions to support stem cell self-renewal and capacity for differentiation. For instance, capillaries can be found in close proximity to cells of the SVZ and hippocampus. Secreted factors and proteins, such as instructive growth factors, regulate neural stem cell behavior and may be absorbed from SVZ capillaries. Such growth factors include epidermal growth factor (EGF), brain-derived neurotrophic factor (BDNF), and vascular endothelial growth factor C (VEGFC), among others. Gliomas are highly vascularized tumors and many tumors overexpress receptors to these growth factors. Furthermore, within the tumor, $CD133^{+}$;$nestin^{+}$ glioma stem cells reside in close proximity to the vasculature [1]. In 3D culture systems, glioma stem cells preferentially home to areas of vasculature, while other cells of the glioma tumor do not display such preference. The factors provided by the vascular endothelial cells were shown to contribute to the maintenance of glioma stem cell proliferation and self-renewal thus explaining their homing preference. Therefore, a strong association exists between vascular and neurogenic niches in adult germinal regions [2, 3]. Vascular endothelial growth factor (VEGF) and its receptors are expressed by neurospheres derived from rodent SVZ [4]. VEGF has also been implicated in glioma growth and is secreted by glioma cells that act on tumor endothelial cells expressing VEGF receptors [5]. Similarly, cancer stem cells isolated from gliomas generate markedly elevated levels of VEGF [6]. Taken together, these data suggest that targeting proangiogenic factors is a potential therapeutic strategy against gliomas and their putative cancer stem cell fraction.

The extracellular matrix (ECM) of germinal regions must contain specialized molecules in order to regulate neuronal differentiation and proper development. The ECM of the SVZ germinal niche is enriched in ECM proteins such as tenascin, collagen, and chondroitin sulfate proteoglycans. Tenascin C is an extracellular matrix molecule that modulates cellular adhesion [7]. The SVZ has increased expression of tenascin C, and it acts to regulate neural stem cell development through the modulation of cell–matrix interactions [8]. Tenascin expression, while concentrated in the human SVZ, is upregulated in GBM and able to stimulate the proliferation of glioma tumor cells. Furthermore, tenascin expression leads to changes in glioma cell gene expression resulting in a more aggressive phenotype [7].

CD44 is a glycoprotein transmembrane receptor and has been used to select for glioma stem cells [9, 10]. It functions as an ECM adhesion protein and its expression is important for non-neoplastic stem cell niche homing and maintenance [11]. CD44 expression in glioma is correlated with more aggressive tumor growth [12] and its expression increases glioma cell migration and invasion [13].

Nestin is a member of a class of intermediate filaments (class VI) that is expressed by neural progenitors during development [14]. It is widely expressed in the brain at birth, but its expression is downregulated in the adult brain and becomes restricted to the SVZ. Nestin^{+} cells also exist in human gliomas [15], lending further support to the hypothesis that neural stem cells may be implicated in glioma formation. Furthermore, nestin expression appears to be significantly correlated to high-grade gliomas in addition to its expression being a predictor for reduced overall survival [16].

Transcription factors play an important role in the regulation of cell fate and are capable of inducing transcriptional programs leading to oncogenesis. The hedgehog family of regulatory pathways is a key regulator of nonneoplastic progenitor proliferation in the SVZ, where the Shh-Gli pathways maintain the stem cell population and facilitate the survival and proliferation of stem cell progeny. It is important to note that Gli is expressed in both low-grade and high-grade gliomas, and that the Shh-Gli pathway may mediate the initiation and maintenance of these tumors as it does for neural stem cells [17]. As might be expected, treatment with cyclopamine (a specific inhibitor of hedgehog signaling) can inhibit the growth of some glioma cell lines in vitro. Thus, the hedgehog family of signaling pathways is implicated in both gliomagenesis and regulation of adult neural stem cell proliferation.

Beyond ECM proteins and transcription factors, growth factor signaling pathways also play an important role in both gliomagenesis and germinal zone regulation. Nearly half of high-grade astrocytomas demonstrate EGF receptor amplification. Not surprisingly, EGFR amplification is a potential transformation mechanism in the development of glioblastoma multiforme. EGF-responsive C cells within the SVZ constitute a large population of migratory, rapidly dividing progenitor. EGF-mediated stimulation prevents C cell differentiation of these cell types and releases their infiltrative potential, similar to the infiltration seen in high-grade gliomas [18].

A population of platelet-derived growth factor (PDGF)+ B cells in the adult SVZ have been identified and shown to give rise to both neurons and oligodendrocytes in vivo [19]. Excessive PDGF activation in the rodent SVZ arrests neuroblast production, induces SVZ cellular proliferation, and creates areas of hyperplasia with features of early glioma formation. There also appears to be a link between these PDGFR+ B cells and the early changes associated with tumor initiation, suggesting that they may be targets of neoplastic transformation.

Models of SVZ Stem Cell Transformation

The genetic manipulation of various known oncogenes, specifically in mouse neural stem cells, has led to the development of numerous animal models of glioma. These models demonstrate the potential for neural stem cell transformation, which often results in neuro-oncogenesis within SVZ cells. Such studies have led to the conclusion that neural stem cell populations are more sensitive to chemical or viral oncogenesis than are areas with a low proportion of proliferating cells. Early models transformation in cells of the SVZ investigated the effects of prenatal exposure to the classic mutagen *n*-ethyl-*n*-nitrosourea (ENU), a well characterized neurocarcinogen [20]. These studies found that cells of the SVZ, in particular, undergo genetic transformations resulting in increased proliferation and immortalization [21]. Interestingly, tumor formation in this model is limited to regions of the SVZ. Genetic mutations in SVZ cells that resulted from ENU exposure included the deletion of INK4a/ARF, cell-cycle genes which are have also been shown to be significantly mutated in GBM samples. The Cancer Genome Atlas 2008 examination of over 200 GBM patient samples found a homozygous deletion or mutation in 49 and 52 % of patients for ARF and INK4a, respectively [22]. Neoplastic cells of mice treated with ENU were also found to upregulate nestin, whose expression is limited to neuronal precursors [23], thus further implicating the role of neural stem cell transformation in the development of glioma.

In addition, an investigation of the loss of tumor suppressor p53 results in increased proliferation of relatively quiescent astrocyte-like SVZ type B cells both in vitro and in vivo [24–26]. Tp53 is a tumor suppressor protein whose signaling is altered in 87 % of GBM patients [22]. When $p53^{-/-}$ mice are prenatally treated with the mutagen ENU, the result is the development glioblastoma-like tumors in 60 % of mice. These tumors form periventricularly and display glioblastoma characteristics, including infiltration into surrounding areas, areas of necrosis, and heterogeneous cell populations. Furthermore, early inactivation of p53 has also been shown to cooperate with the neurofibromatosis-1 (NF1) tumor suppressor gene mutation, resulting in malignant astrocytoma formation in a mouse tumor model [27, 28]. The NF1 tumor suppressor neurofibromin is a functional Ras GTPase-activating protein and its loss results in abnormal activation of Ras, a central mediator of receptor tyrosine kinase (RTK) signaling. Furthermore, mutation or homozygous deletion in NF1 has been noted in 18 % of GBM samples [22]. Mice that carry germline mutations in both p53 and NF1 develop both low and intermediate-grade astrocytomas. These astrocytomas express Nestin, the progenitor-associated intermediate filament, and were consistently associated with the SVZ. Based on these results, it appears that SVZ cells are most susceptible to p53/NF1-mediated astrocytoma formation and that the cell-of-origin for malignant astrocytomas in p53/NF1 mutant mice may reside within the SVZ.

Various other mouse models of neural stem cell transformation have provided clues as to the potential molecular events that lead to neuro-oncogenesis [27, 29–33]. The genetic manipulation of Harvey-Ras (H-Ras) and AKT in as little as 60 $GFAP^+$ precursor cells of the SVZ or hippocampus results in the development of high-grade gliomas in these regions [33]. Both the H-Ras pathway and the AKT pathway are highly associated with gliomas. Recent studies have reported 2 % of gliomas experience gain-of-function mutations in Ras and 36 % of gliomas experience a loss of the tumor suppressor phosphatase and tensin homolog (PTEN), a negative regulator of the Ras signaling pathway. Both a gain-of-function mutation in Ras and loss-of-function mutation in PTEN result in AKT pathway activation. These pathways participate in RTK signaling, which has been shown to be altered in 88 % of GBM patients [22]. Importantly, the injection of active AKT and H-RasV12 into the cortex fails to result in any significant tumor formation. However, mice with injection in the hippocampus or SVZ results in neuro-oncogenesis of glioma-like tumors, indicating $GFAP^+$ NSCs of the SVZ and hippocampus to be putative cells of origin [33]. These tumors display pathological characteristics of glioblastomas, including microvascular proliferation, pseudo-palisading necrosis, and increased cell density. In addition to displaying these hallmark characteristics of glioblastoma, the tumors also exhibit cellular heterogeneity, evidenced by the expression of various cellular markers, including the astrocytic marker GFAP, the oligodendrocyte marker, myelin basic protein, and a neuronal specific marker, tuj1. Furthermore, the tumors in the hippocampus and/or SVZ were mostly GFP^+, indicating that a majority of the cells in the tumor were derived from a smaller infected cell population. After isolating GFP^+ cells from the tumors of GFAP-Cre;$TP53^{+/-}$ mice injected with H-RasV12/AKT and culturing them in neural stem cell media, the cells formed neurosphere structures in vitro. Proliferation and differentiation assays confirmed that these cells were both self-renewing and multipotent, as they were proliferative in neural stem cell media and differentiated upon serum stimulation. Forty to fifty percent of these cells express CD133, a speculative marker for glioma initiating cells. This study provides strong evidence towards the SVZ neural stem cell as the putative cell-of-origin for GBM as tumors arise in the SVZ and hippocampus and fail to develop in the cortex. Furthermore, these glioma initiating cells of the SVZ maintained their self-renewal and differentiation capacities, indicating that stemness is a contributing factor to astrocytoma development.

PTEN (phosphatase and tensin homologue) is a recognized tumor suppressor mutated or deleted in 36 % of GBM samples [22]. PTEN protein is a phosphatidylinositol phosphate (PIP) phosphatase that lowers PIP3 levels and enhances the rate of apoptosis. PTEN also decreases cell motility via G protein-coupled mechanisms. PTEN is expressed in SVZ precursor cells during neuronal differentiation [34]. A loss of PTEN and p53 in neural stem cells isolated from the SVZ results in significantly increased proliferation, self-renewal capacity, and impaired differentiation potential [35]. In vivo, 42 out of 57 hGFAP-Cre^{+};p53$^{fl/fl}$;Pten$^{fl/+}$ mice develop malignant gliomas which display classic features of human GBM including cellular pleomorphism, microvascular proliferation, and areas of necrosis. In another mouse model, heterozygous loss of FN1 and PTEN coupled with homozygous loss of p53 in nestin-cre^{+} cells results in 100 % astrocytoma tumor development [32]. Again, it is important to note that viral injection of cre into the SVZ of these mice results in astrocytoma formation in 100 % of mice, while injection of cre into other regions, such as the cortex and striatum, does not. Studies such as these, which uncover the innate tumor competence of SVZ neural stem cells, provide strong evidence for the neural stem cell as the cell-of-origin for glioma.

Clinical and Therapeutic Implications

Clinical correlations have been drawn between GBMs and their anatomical relationship to the SVZ [36–38]. An examination of 53 patients with newly diagnosed GBM and the classification of their tumors based on spatial relationship to the SVZ showed that patients whose tumors made contact with the SVZ had a significantly higher incidence of multifocal disease at diagnosis [37]. Interestingly, upon follow-up MRI, all of the patients whose tumors had direct contact to the SVZ had recurred in a noncontiguous manner with the original lesion. However, of the patients whose tumors did not have contact with the SVZ, none had any evidence of tumor noncontiguous with the primary lesion [37]. A similar analysis of 91 GBM patients showed that progression free survival is significantly reduced in patients whose tumor has contact with both the SVZ and the cortex [36]. Furthermore, those patients with SVZ involvement had an overall reduced survival as well as a decreased time until recurrence compared to those tumors not involving the SVZ. In addition, an analysis of 39 newly diagnosed GBM patients revealed that tumor SVZ involvement was a significant predictor in reduced overall survival [38].

Based on the cancer stem cell theory, any brain tumor therapy that fails to eradicate cancer stem cells will result in recurrence or regrowth of the residual tumor stem cells, resulting in eventual disease progression [39]. Recently, studies have attempted to ask whether irradiation of the ipsilateral SVZ can improve GBM prognosis [40–45]. The first of such studies analyzed 55 patients with glioma who received surgery, radiotherapy and chemotherapy [42]. SVZ radiation dose was found to be significantly predictive of progression free survival, and bilateral radiation to the SVZ yielded a significant hazard ratio for death, leading to the author's conclusion that SVZ radiation may provide a significant benefit for GBM survival. In a pooled analysis of 173 GBM patients from two academic centers, these results were confirmed as high radiation therapy doses to the ipsilateral SVZ led to significantly longer progression free survival. An analysis of 40 GBM patients found statically significant improved overall survival in patients who received high dose of ipsilateral SVZ radiation [43]. In a recent investigation of 100 GBM patients—50 long term survivors (>3 years) and 50 short term survivors (<1 year), SVZ tumor contact was found to be a significant predictor of prolonged survival, in addition to age and total resection status [40].

While these data are suggestive, the topic of the neural stem cell as the cell-of-origin for glioma remains controversial. Other hypotheses of glioma initiating cells exist, including the notion of cellular dedifferentiation resulting in neuro-oncogenesis [46]. In this model, terminally differentiated cells within the tumor are capable of dedifferentiating and reverting back to multipotency. Others have shown differentiated terminal astrocytes, in addition to NSCs, are capable of dedifferentiation under the influence of Ink4a/Arf inactivation and EGFR activation [47]. Furthermore, the oligodendrocyte precursor cell (OPC) is also postulated as a candidate cell of origin for glioma [48–50]. Using mosaic analysis with double markers (MADM) in mice, Liu et al. [49] show that only OPCs are capable of gliomagenesis upon genetic manipulation. Yet another OPC cell of origin model shows $NG2^+$ oligodendrocytes may undergo a loss of asymmetric divisions resulting in proliferative, self-renewing cells with tumor-initiating potential [50]. Moving forward, increased focus on the development of these brain tumors, including cellular and molecular transformations in different cell types, may lead to the discovery of new therapeutic targets able to arrest neuro-oncogenesis early in its track.

Conclusions

The discovery of neural stem cells in the adult human brain has led to the emergence of a new area of scientific inquiry connecting neuro-oncology with developmental neurobiology. This field has gained prominence in recent years with the identification and characterization of stem-like cells within glioma tumors which retain the capacity to both self-renew and differentiate into neuronal subtypes. Glioma tumors have been shown to resemble neuronal germinal niches in structural, functional, and molecular characteristics. Findings such as these may pave the way for the identification of a cell-of-origin for human glioma allowing for the development of novel therapeutic agents and strategies to improve glioma prognosis.

References

1. Calabrese C, Poppleton H, Kocak M, Hogg TL, Fuller C, Hamner B, Allen M. A perivascular niche for brain tumor stem cells. Cancer Cell. 2007;11(1):69–82.
2. Gilbertson RJ, Rich JN. Making a tumour's bed: glioblastoma stem cells and the vascular niche. Nat Rev Cancer. 2007;7(10):733–6.
3. Palmer TD, Willhoite AR, Gage FH. Vascular niche for adult hippocampal neurogenesis. J Comp Neurol. 2000;425(4):479–94.
4. Knizetova P, Darling JL, Bartek J. Vascular endothelial growth factor in astroglioma stem cell biology and response to therapy. J Cell Mol Med. 2008;12(1):111–25.
5. Jain RK, Di Tomaso E, Duda DG, Loeffler JS, Sorensen AG, Batchelor TT. Angiogenesis in brain tumours. Nat Rev Neurosci. 2007;8(8):610–22.

6. Bao S, Wu Q, Sathornsumetee S, Hao Y, Li Z, Hjelmeland AB, Rich JN. Stem cell-like glioma cells promote tumor angiogenesis through vascular endothelial growth factor. Cancer Res. 2006;66(16):7843–8.
7. Ruiz C, Huang W, Hegi ME, Lange K, Hamou M-F, Fluri E, Orend G. Differential gene expression analysis reveals activation of growth promoting signaling pathways by tenascin-C. Cancer Res. 2004;64(20):7377–85.
8. Garcion E, Halilagic A, Faissner A. Generation of an environmental niche for neural stem cell development by the extracellular matrix molecule tenascin C. Development. 2004;131(14): 3423–32.
9. Anido J, Sáez-Borderías A, Gonzàlez-Juncà A, Rodón L, Folch G, Carmona MA, Prudkin L. TGF-β Receptor Inhibitors Target the CD44 ^{high}/Id1 ^{high} Glioma-Initiating Cell Population in Human Glioblastoma. Cancer Cell. 2010;18(6):655–68.
10. Fu J, Yang Q-Y, Sai K, Chen F-R, Pang JC, Ng H-K, Chen Z-P. TGM2 inhibition attenuates ID1 expression in CD44-high glioma-initiating cells. Neuro Oncol. 2013;15(10):1353–65.
11. Haylock DN, Nilsson SK. Perspective stem cell regulation by the hematopoietic stem cell niche. Cell Cycle. 2005;4(10):1353–5.
12. Pietras A, Katz AM, Ekström EJ, Wee B, Halliday JJ, Pitter KL, Holland EC. Osteopontin-CD44 signaling in the glioma perivascular niche enhances cancer stem cell phenotypes and promotes aggressive tumor growth. Cell Stem Cell. 2014;14(3):357–69.
13. Chetty C, Vanamala SK, Gondi CS, Dinh DH, Gujrati M, Rao JS. MMP-9 induces CD44 cleavage and CD44 mediated cell migration in glioblastoma xenograft cells. Cell Signal. 2012;24(2):549–59.
14. Wiese C, Rolletschek A, Kania G, Blyszczuk P, Tarasov K, Tarasova Y, Wobus A. Nestin expression: a property of multi-lineage progenitor cells? Cell Mol Life Sci CMLS. 2004;61(19–20):2510–22.
15. Yang XH et al. Nestin expression in different tumours and its relevance to malignant grade. J Clin Pathol. 2008;61(4):467–473.
16. Arai H, Ikota H, Sugawara K-I, Nobusawa S, Hirato J, Nakazato Y. Nestin expression in brain tumors: its utility for pathological diagnosis and correlation with the prognosis of high-grade gliomas. Brain Tumor Pathol. 2012;29(3):160–7.
17. Dahmane N, Sánchez P, Gitton Y, Palma V, Sun T, Beyna M, Ruiz i Altaba A. The Sonic Hedgehog-Gli pathway regulates dorsal brain growth and tumorigenesis. Development. 2001;128(24):5201–12.
18. Doetsch F, Petreanu L, Caille I, Garcia-Verdugo J-M, Alvarez-Buylla A. EGF converts transit-amplifying neurogenic precursors in the adult brain into multipotent stem cells. Neuron. 2002;36(6):1021–34.
19. Jackson EL, Garcia-Verdugo JM, Gil-Perotin S, Roy M, Quinones-Hinojosa A, VandenBerg S, Alvarez-Buylla A. PDGFRα-positive B cells are neural stem cells in the adult SVZ that form glioma-like growths in response to increased PDGF signaling. Neuron. 2006;51(2):187–99.
20. Sanai N. Adult neural stem cells and gliomagenesis. In: Glioblastoma. Springer; 2010. p. 153–65.
21. Savarese TM, Jang T, Pang Low H, Salmonsen R, Litofsky NS, Matijasevic Z, Recht LD. Isolation of immortalized, INK4a/ARF-deficient cells from the subventricular zone after in utero N-ethyl-N-nitrosourea exposure. J Neurosurg. 2005;102(1):98–108.
22. McLendon R, Friedman A, Bigner D, Van Meir EG, Brat DJ, Mastrogianakis GM, Aldape K. Comprehensive genomic characterization defines human glioblastoma genes and core pathways. Nature. 2008;455(7216):1061–8.
23. Jang T, Litofsky NS, Smith TW, Ross AH, Recht LD. Aberrant nestin expression during ethylnitrosourea-(ENU)-induced neurocarcinogenesis. Neurobiol Dis. 2004;15(3):544–52.
24. Gil-Perotin S, Marin-Husstege M, Li J, Soriano-Navarro M, Zindy F, Roussel MF, Casaccia-Bonnefil P. Loss of p53 induces changes in the behavior of subventricular zone cells: implication for the genesis of glial tumors. J Neurosci. 2006;26(4):1107–16.

25. Katayama K-I, Ueno M, Yamauchi H, Nagata T, Nakayama H, Doi K. Ethylnitrosourea induces neural progenitor cell apoptosis after S-phase accumulation in a p53-dependent manner. Neurobiol Dis. 2005;18(1):218–25.
26. Leonard JR, D'Sa C, Klocke BJ, Roth KA. Neural precursor cell apoptosis and glial tumorigenesis following transplacental ethyl-nitrosourea exposure. Oncogene. 2001;20(57): 8281–6.
27. Wang Y, Yang J, Zheng H, Tomasek GJ, Zhang P, McKeever PE, Zhu Y. Expression of mutant p53 proteins implicates a lineage relationship between neural stem cells and malignant astrocytic glioma in a murine model. Cancer Cell. 2009;15(6):514–26.
28. Zhu Y, Guignard F, Zhao D, Liu L, Burns DK, Mason RP, Parada LF. Early inactivation of p53 tumor suppressor gene cooperating with NF1 loss induces malignant astrocytoma. Cancer Cell. 2005;8(2):119–30.
29. Chen J, McKay RM, Parada LF. Malignant glioma: lessons from genomics, mouse models, and stem cells. Cell. 2012;149(1):36–47.
30. Chow LM, Endersby R, Zhu X, Rankin S, Qu C, Zhang J, Baker SJ. Cooperativity within and among Pten, p53, and Rb pathways induces high-grade astrocytoma in adult brain. Cancer Cell. 2011;19(3):305–16.
31. Lei L, Sonabend AM, Guarnieri P, Soderquist C, Ludwig T, Rosenfeld S, Canoll P. Glioblastoma models reveal the connection between adult glial progenitors and the proneural phenotype. PLoS One. 2011;6(5):e20041.
32. Llaguno SRA, Chen J, Parada LF. Signaling in malignant astrocytomas: role of neural stem cells and its therapeutic implications. Clin Cancer Res. 2009;15(23):7124–9.
33. Marumoto T, Tashiro A, Friedmann-Morvinski D, Scadeng M, Soda Y, Gage FH, Verma IM. Development of a novel mouse glioma model using lentiviral vectors. Nat Med. 2009;15(1):110–6.
34. Li L, Liu F, Salmonsen RA, Turner TK, Litofsky NS, Di Cristofano A, Ross AH. PTEN in neural precursor cells: regulation of migration, apoptosis, and proliferation. Mol Cell Neurosci. 2002;20(1):21–9.
35. Zheng H, Ying H, Yan H, Kimmelman AC, Hiller DJ, Chen A-J, Ding Z. p53 and Pten control neural and glioma stem/progenitor cell renewal and differentiation. Nature. 2008;455(7216): 1129–33.
36. Jafri NF, Clarke JL, Weinberg V, Barani IJ, Cha S. Relationship of glioblastoma multiforme to the subventricular zone is associated with survival. Neuro Oncol. 2013;15(1):91–6.
37. Lim DA, Cha S, Mayo MC, Chen M-H, Keles E, VandenBerg S, Berger MS. Relationship of glioblastoma multiforme to neural stem cell regions predicts invasive and multifocal tumor phenotype. Neuro Oncol. 2007;9(4):424–9.
38. Young GS, Macklin EA, Setayesh K, Lawson JD, Wen PY, Norden AD, Kesari S. Longitudinal MRI evidence for decreased survival among periventricular glioblastoma. J Neurooncol. 2011;104(1):261–9.
39. Schulenburg A, Ulrich-Pur H, Thurnher D, Erovic B, Florian S, Sperr WR, Zielinski CC. Neoplastic stem cells: a novel therapeutic target in clinical oncology. Cancer. 2006; 107(10):2512–20.
40. Adeberg S, Bostel T, König L, Welzel T, Debus J, Combs SE. A comparison of long-term survivors and short-term survivors with glioblastoma, subventricular zone involvement: a predictive factor for survival? Age (years). 2014;64:8.
41. Corn BW, Raizer J, Kanner AA. Should the subventricular zone be part of the "rad" zone? J Neurooncol. 2014;118:423–4.
42. Evers P, Lee PP, DeMarco J, Agazaryan N, Sayre JW, Selch M, Pajonk F. Irradiation of the potential cancer stem cell niches in the adult brain improves progression-free survival of patients with malignant glioma. BMC Cancer. 2010;10(1):384.
43. Gupta T, Nair V, Paul SN, Kannan S, Moiyadi A, Epari S, Jalali R. Can irradiation of potential cancer stem-cell niche in the subventricular zone influence survival in patients with newly diagnosed glioblastoma? J Neurooncol. 2012;109(1):195–203.

44. Lee P, Eppinga W, Lagerwaard F, Cloughesy T, Slotman B, Nghiemphu PL, Demarco J. Evaluation of high ipsilateral subventricular zone radiation therapy dose in glioblastoma: a pooled analysis. Int J Radiat Oncol Biol Phys. 2013;86(4):609–15.
45. Sonoda Y, Saito R, Kanamori M, Kumabe T, Uenohara H, Tominaga T. The association of subventricular zone involvement at recurrence with survival after repeat surgery in patients with recurrent glioblastoma. Neurol Med Chir. 2014;54(4):302–9.
46. Friedmann-Morvinski D, Bushong EA, Ke E, Soda Y, Marumoto T, Singer O, Verma IM. Dedifferentiation of neurons and astrocytes by oncogenes can induce gliomas in mice. Science. 2012;338(6110):1080–4.
47. Bachoo RM, Maher EA, Ligon KL, Sharpless NE, Chan SS, You MJ, Weissleder R. Epidermal growth factor receptor and *Ink4a/Arf*: Convergent mechanisms governing terminal differentiation and transformation along the neural stem cell to astrocyte axis. Cancer Cell. 2002;1(3):269–77.
48. Lindberg N, Kastemar M, Olofsson T, Smits A, Uhrbom L. Oligodendrocyte progenitor cells can act as cell of origin for experimental glioma. Oncogene. 2009;28(23):2266–75.
49. Liu C, Sage JC, Miller MR, Verhaak RG, Hippenmeyer S, Vogel H, Luo L. Mosaic analysis with double markers reveals tumor cell of origin in glioma. Cell. 2011;146(2):209–21.
50. Sugiarto S, Persson AI, Munoz EG, Waldhuber M, Lamagna C, Andor N, Siu J. Asymmetry-defective oligodendrocyte progenitors are glioma precursors. Cancer Cell. 2011;20(3):328–40.

Isolation and Characterization of Stem Cells from Human Central Nervous System Malignancies

Imad Saeed Khan and Moneeb Ehtesham

Abstract Central Nervous System (CNS) tumors include some of the most invasive and lethal tumors in humans. The poor prognosis in patients with CNS tumors is ascribed to their invasive nature. After the description of a stem cell-like cohort in hematopoietic cancers, tumor stem cells (TSCs) have been isolated from a variety of solid tumors, including brain tumors. Further research has uncovered the crucial role these cells play in the initiation and propagation of brain tumors. More importantly, TSCs have also been shown to be relatively resistant to conventional cytotoxic therapeutics, which may also account for the alarmingly high rate of CNS tumor recurrence. In order to elucidate prospective therapeutic targets it is imperative to study these cells in detail and to accomplish this, we need to be able to reliably isolate and characterize these cells. This chapter will therefore, provide an overview of the methods used to isolate and characterize stem cells from human CNS malignancies.

Keywords Glioma stem cells • Tumor stem cells • Cancer stem cells • Stem cell sorting • Hoechst dye exclusion • ALDH1 assay • Neurosphere culture

Introduction

Tumors of the central nervous system (CNS) include some of the most lethal malignancies. While there has been sizeable development in the management strategies used to combat intrinsic CNS tumors there is still room for improvement. Glioblastoma Mutliforme (GBM) is the most common kind of primary brain tumor and carries a poor prognosis. With optimized surgical resection combined with chemotherapy and radiotherapy, the median survival rate is approximately 14 months [1].

I.S. Khan, M.D.
Section of Neurosurgery, Geisel School of Medicine at Dartmouth, Lebanon, NH, USA

M. Ehtesham, M.D. (✉)
Department of Neurological Surgery, Vanderbilt University Medical Center, Nashville, TN, USA
e-mail: moneeb.ehtesham@Vanderbilt.Edu

M. Ehtesham (ed.), *Stem Cell Biology in Neoplasms of the Central Nervous System*, Advances in Experimental Medicine and Biology 853,
DOI 10.1007/978-3-319-16537-0_3

Due to the very invasive nature of the tumor, most of the treated patients eventually succumb to the disease.

Likewise, the most common brain tumor in pediatric patients is medulloblastoma. Treatment consists of maximal surgical resection followed by chemo-radiation and the median survival rate is a little more than 5 years. However, patients presenting with more advanced disseminated disease fare much worse. Furthermore, treated patients face a long-term prognosis that is fraught with increased risk of secondary malignances and cognitive deficits.

The poor prognosis of patients with CNS tumors along with the near absence of treatment modalities that have improved outcome significantly has lead to investigators looking at the biology of the tumors more closely. This has steered us to the identification of a small subset of tumor cells that have stem cell like properties of cell renewal and lineage capacity [2, 3]. These tumor stem cells (TSCs) have been identified as the key mediators of tumor initiation, propagation and maintenance [4]. Moreover, recent evidence suggests that TSCs play an important part in angiogenesis and are relatively refractory to conventional chemo-radiation therapeutics [2, 5–8]. It is now thought that these cells may play an integral role in the recurrence of CNS tumors [9, 10].

Preparation for TSC Isolation

To be able to therapeutically target TSCs, it is imperative to be able to isolate and study their genetic and proteomic characteristics. Isolation of TSCs is a challenge because of their seemingly small number and the fact that they share some of the properties of downstream committed progenitor cells, which may also be able to repopulate the tumor [2]. The first step in isolating TSCs is to obtain a single-cell suspension of the tumor. This may be achieved with mechanical dissociation, chemical dissociation, or a combination of both.

To create a single-cell suspension it is important that the tumor be processed as soon as possible after the excision to retain viability. This may be a challenge in tumors excised from patients but is suggested that tissue processing should begin within 30 min of the removal. The tumor sample is then placed in the sterile Petri dish with HBSS+Ca+Mg, and using a dissecting microscope cleaned off all the necrotic tissue and blood vessels. The sample can then be mechanically dissociated with microscissors or the unsharpened side of a #22 scalpel blade. The cell slurry created is then dissociated using pipettes and passing through a 70 μm filter. Alternatively, Trypsin–EDTA may be added to the clean sample to help dissociation. The decision to carry out mechanical vs. chemical dissociation is subject to investigator preference and the type of tumor used (for detailed methods see refs. [11, 12]).

Once the solid CNS tumor has been dissociated into a single-cell suspension, isolation of TSCs can be accomplished by three major categories: sorting cells based on cell surface markers, choosing cells with the expression of a particular protein of interest, and functional assays. As the recognition of TSCs is based on the multipotency and clonogenicity of the cells in question, TSC isolation methods are best followed by in vivo functional assays to confirm identity [3]. This chapter further provides an overview of some of the most commonly used CNS TSC isolation and characterization methods.

Stem Cell Sorting

The two most important aims of cell sorting techniques are to preserve the viability of the cells and to get the purest selected fraction possible. Flow cytometry is a powerful tool to select cells expressing particular cell surface markers (Fig. 1).

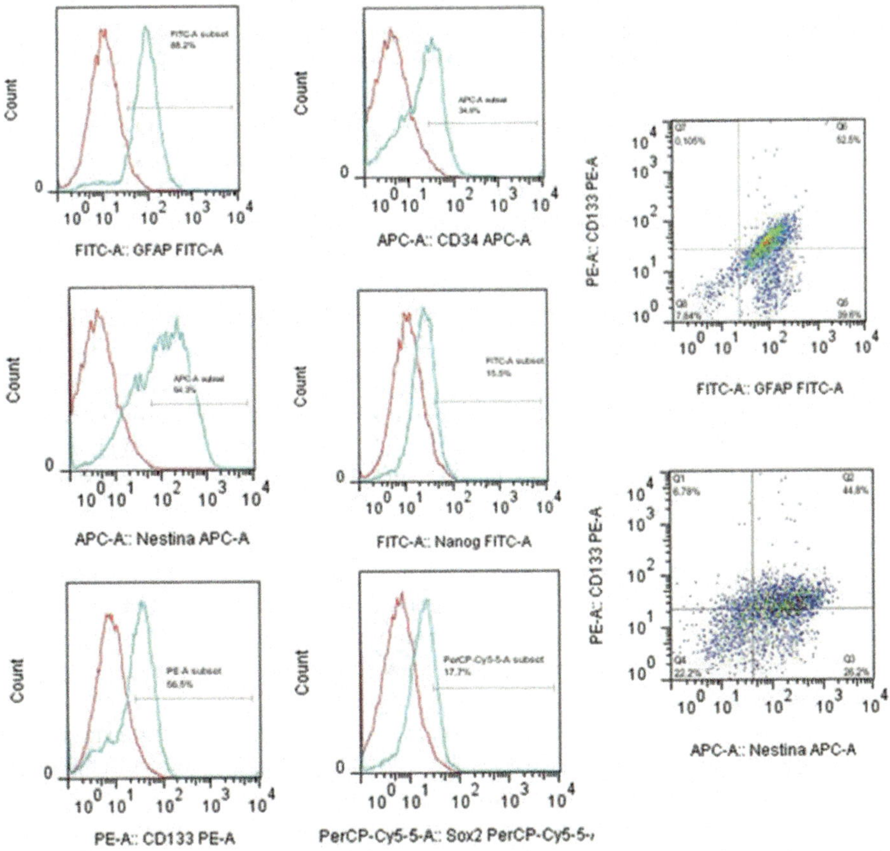

Fig. 1 Immunophenotyping of tumor subspheres by flow cytometry. Immunophenotypic characterization by flow cytometry assays showing the pattern of expression of markers GFAP (88.2 %), CD133 (56.5 %), Nestin (64.2 %), Sox2 (17.7 %), CD34 (34.0 %), and Nanog (15.5 %) in glioblastoma subsphere samples and the co-expression of CD133 with GFAP (52.5 %) and Nestin (44.8 %). Representative figure of five samples of glioblastoma. (From Pavon LF et al. Front Neurol. 2014 Jan 7;4:214. 2014. In vitro Analysis of Neurospheres Derived from Glioblastoma Primary Culture: A Novel Methodology Paradigm (open access))

This technique can be used to separate TSCs from the complex structure of a solid tumor. Another advantage of flow cytometry is the use of multiple markers to simultaneously positively or negatively select out the cells. Suitable cell surface antibodies conjugated to fluorophores (fluorescent chemical compounds that can re-emit light upon light excitation) are selected and then added to the sorting sample. The sample is then passed through the flow cytometer to sort out the cells with the surface marker of interest.

In the context of cell surface markers, CD133 (Prominin-1) is one of the most commonly utilized markers to identify neural stem cells (NSC) and TSC [13]. CD133 is a pentaspan transmembrane glycoprotein, which localizes on cellular protrusions [14]. Weigman and colleagues initially identified the glycoprotein by raising monoclonal antibodies against mouse neuroepithelium [15]. Around the same time, Yin et al. and Miraglia et al. also identified CD133 independently, using antibodies against $CD34^{+}$ hematopoietic stem and progenitor cells [16, 17].

CD133 was initially used to enrich cancer stem cells in leukemia [18, 19], and has been observed in various other tumor stem cells from different cancers (review in [20]). Owing to its presence on NST, Singh and colleagues conducted in vitro and in vivo studies showing the presence of TSC in CD133+ cells in gliomas and medulloblastomas [21, 22]. Along with CD133, neurospheres derived from pediatric brain tumors may also express other surface markers, including Sox2, musashi-1, bmi-1, maternal embryonic leucine zipper kinase, and phosphoserine phosphatase [23].

Nestin is another important marker for CNS TSCs [24]. Initially described as an antigen of RAT401 against embryonic spinal cord, Nestin was later identified as a class VI intermediate filament protein [25, 26]. Alongside CD133, Nestin has been shown to correlate with the aggressiveness of gliomas in some studies [27, 28].

It is important to note that one of major issues facing stem cell surface markers is the fact that they are also found on other cell populations [29]. For instance, Nestin is also expressed by Bergmann glia and granule neuron precursors in the cerebellum [30, 31]. Similarly, CD133 has been observed in mature astrocytes, oligodendrogliocytes, and neurons, as well as ependymal cells [32, 33].

Magnetic bead separation is an alternative to flow cytometry and allows the cells to be separated by incubating magnetic nanoparticles coated with antibodies against a particular surface antigen/marker. The nanoparticles attach to the cell surface marker of interest and the sample is then sorted into marker positive and negative groups by flowing the cells through a strong magnetic field.

The use of magnetic bead separation for TSC has been described in a variety of studies [34–36]. As this modality passes the cells as a group, in contrast to the flow cytometer that passes cells individually, a larger sample can be separated rapidly. This is of particular advantage in cases where the population of cells positive for the labeled antibody is very small, requiring a large tumor sample to be passed to get an adequate number of sorted cells. However, magnetic bead sorting can only separate one marker at a time and hence multiple runs may be required to sort out cells if multiple markers have to be used. This increases the time used for sorting and may decrease cell viability.

Hoechst Dye Exclusion

An alternative technique used to identify TSCs is by using the Hoechst dye exclusion in tumor cells. Hoechst stains are part of a family of blue fluorescent dyes used to stain DNA [37, 38]. Hoechst dye 33342 can be used in conjunction with flow cytometer to identify the side-population (SP) of cells. Goodell and colleagues were the first to identify a side-population (SP) from mouse bone marrow that was enriched with cells that had properties of stem cells [39]. The same group later isolated stem cells in humans [40].

Since then, several studies have described the use of this technique to isolate TSCs from a wide variety of cancers [41–48]. The unique property of SP cells is that they actively eliminate the dye leading to a low Hoechst staining (dye exclusion). The enhanced dye efflux in the side-population cells is due to an increase in the activity of multi-drug resistance proteins, primarily ABCG2 [49]. To carry out the protocol, Hoechst 33342 is added to the cell suspension to be tested. The cells can then be separated using flow cytometric sorting. Drugs such as verapamil, reserpine, or fumitremorgin C inhibit Hoechst exclusion and abolish the SP. Co-staining with antibodies helps with the identification and confirmation of the cells [11].

Whereas this technique has been useful in isolating stem cell enriched populations from CNS tumors and cell lines, it is important to note that on its own this technique isolates a population enriched for but not homogeneous for TSC. Additionally there is increasing evidence that the Hoechst exclusion alone may not be sufficient to isolate CNS TSCs [41, 42, 50]. Indeed, Broadley et al. found that while neurospheres were able to enrich for TSC from primary GBM cells and GBM cell lines, no SP was found when the neurospheres were analyzed. This led them to conclude that SP was not "necessary or sufficient" for a TSC phenotype in GBM [51].

ALDH1 Assay

Aldehyde dehydrogenases are a large group of enzymes that catalyze the oxidation of aldehydes [52], and are found in the cytosol, nucleus, mitochondria, and endoplasmic reticulum [53]. Of the ALDH group, the ALDH enzymes associated with NSC and TSC include ALDH1, ALDH2*2, ALDH3A1, ALDH4A1, and ALDH7A1 [52]. Of these, ALDH1 and ALDH3A1 in particular have been shown to play an important functional role in TSC. ALDH1 is critical in the Retinoid signaling pathway, which plays an important role in regulation of gene expression, morphogenesis and development of NSC and TSC [54–56]. ALDH1A1 and ALDH3A1 have also been shown to offer protection against alkylating agents, such as cyclophosphamide [57, 58].

Jones and colleagues were the first group to report a method to measure the intracellular ALDH1 activity in viable cells [59]. They used dansyl aminoacetaldehyde

(DAAA), a fluorescent aldehyde, in flow cytometry experiments to isolate viable mouse and human cells based on their ALDH content. However this technique can be mutagenic to the isolated cells as the DAAA fluorescence is excited by UV emissions. Additionally, the emission spectra of DAAA overlap with other fluorochromes, which makes it more challenging to carry out simultaneous analysis of other cell markers [59, 60]. Storms and colleagues modified the technique by using BODIPY aminoacetaldehyde (BAAA)—a fluorescent substrate for ALDH [60]. This assay is also known as the Aldefluor® Assay and has been shown to be useful method to isolate NSCs and TSCs without any ex vivo manipulation of the cells [61].

In the context of gliomas, Rasper et al. showed that high protein levels of ALDH1 facilitate neurosphere formation in established GBM cell lines, and even single ALDH1 positive cells could give rise to neurospheres [62]. Mao et al. characterized two mutually exclusive glioma TSC subtypes, and showed that ALDH activity was significantly elevated in Mesenchymal (Mes) TSCs but not in Proneural (PN) GSCs [63]. Additionally, inhibition of ALDH1A3 attenuated the growth of Mes but not PN GSCs, suggesting that ALDH1A3 pathways are promising therapeutic targets (Fig. 2).

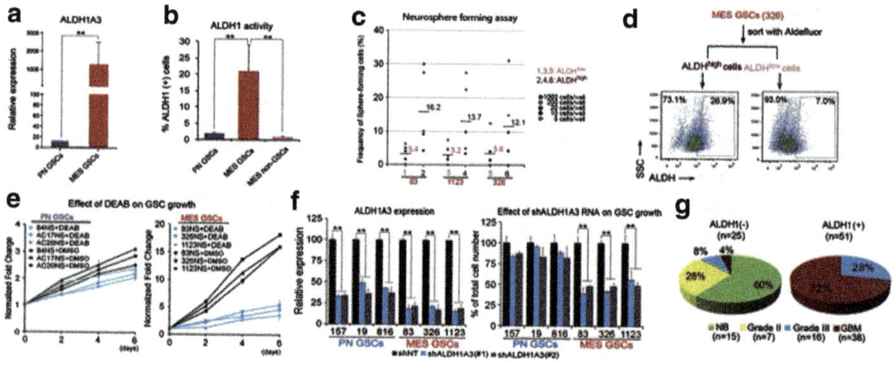

Fig. 2 ALDH1A3 is a functional Mes GSC marker. (**a**) qRT-PCR analysis of ALDH1A3 expression in PN and Mes GSCs (**$P<0.01$). (**b**) FACS analysis using Aldefluor. ALDH activities in PN GSCs ($n=3$), Mes GSCs ($n=3$), and non-GSCs ($n=3$) derived from Mes GSCs (**$P<0.01$). (**c**) Frequency of sphere-forming cells between ALDH1^{high} and ALDH1^{low} Mes GSCs. FACS-sorted based on ALDH expression Mes GSCs were used in the assays (**$P<0.01$). (**d**) FACS reanalysis: ALDH activity after 1-week postcell sorting of Mes 326 ALDH^{high} cells. ALDH^{high} Mes GSC spheres generated both ALDH^{high} and ALDH^{low} cells, whereas the majority of ALDH^{low} sphere cells retain as ALDH^{low} cells. (**e**) Effect of an ALDH inhibitor DEAB on cell growth of PN ($n=3$) and Mes ($n=3$) GSCs. DEAB abrogates the in vitro growth of Mes GSCs but has a marginal effect on PN GSCs. (**f**) Effect of shALDH1A3 knockdown on growth and ALDH1A3 gene expression of both PN and Mes GSCs. The growth of Mes GSCs is significantly reduced by shRNA-mediated depletion of ALDH1A3 compared with PN GSCs. RNA interference with 2 shALDH1A3 constructs significantly reduced ALDH1A3 expression levels in PN and Mes GSCs ($n=3$ each, **$P<0.01$). (**g**) Pie chart indicating the number of samples that were analyzed in different WHO tumor grades of clinical glioma samples or normal brain tissues that are ALDH(+) or (−). Data in (**a–f**) are representative of three independent experiments with similar results. (With permission from Mao et al. Mesenchymal glioma stem cells are maintained by activated glycolytic metabolism involving aldehyde dehydrogenase 1A3. Proc Natl Acad Sci U S A. 2013 May 21;110(21):8644–9)

ALDH1 levels have also been correlated with outcome in patients with glioma. For instance, ALDH1A3 promoter methylation has been found to confer a favorable prognosis for patients with GBMs [64]. Liu and colleagues analyzed their samples of astrocytomas and reported an association of ALDH1 expression with pathological grade and patient survival [65].

Dye Based Isolation

The identification of a dye-retaining brain tumor population can also enable the identification of a subpopulation displaying the hallmarks of TSC. Label-retaining cell fractions that enrich TSCs have been isolated from various solid tumors, including breast [66], pancreatic [67], and skin tumors [68].

In the context of CNS tumors, Deleyrolle and colleagues used the properties of the pro-drug carboxyfluorescein diacetate succinimidylester (CFSE), which is converted by cellular esterase activity into a fluorescent compound covalently bound to proteins and retained within the cells [69]. CFSE dye can enable quantification of cell proliferation, as it is equally divided between daughter cells after division. They observed that a sub-population of cells diluted the dye significantly slower than the overall population, presumably due to a lower frequency of cell division. Using a limiting dilution transplantation assay in immunocompromised mice, these label-retaining brain tumor cells displayed elevated tumor-initiation properties relative to the bulk population, depicting their stem cell-like properties [69].

Spectroscopy

Other novel methods of TSC identification based on the biochemical composition of the TSC have also been described. One of these modalities is Fourier transform infrared spectroscopy (FTIR). Vibrational spectroscopy enables the label-free characterization of cells by probing the biochemical composition and numerous groups have described the classification of gliomas using this modality [70–72].

Wehbe and colleagues were able to depict the difference between normal and tumor vasculature of animal and human glioma using FTIR imaging [73]. More recently, Uckermann and colleagues described the identification of GBM TSC using FTIR [74]. They were able to discern biochemical differences between GBM cell populations with high and low content of TSCs that were likely related to differences in the RNA/DNA content.

Neurosphere Culture

While the previously mentioned techniques show great utility in the isolation of TSC, they are limited by their nature of using proxy characteristics to separate the cells. In this setting, neurospheres can help characterize TSC more definitively. This culture method relies upon the ability of TSC to initiate growth, self-renew, and display multi-potentiality through generation of its progeny [75].

Neurospheres refer to in vitro three-dimensional free floating spheroid cellular clusters in a supernatant that form when TSCs are exposed to a serum-free environment. The assay uses epidermal growth factor (EGF) and basic fibroblast growth factor (FGF), and has the property of selectively supporting the growth and proliferation of cells with stem cell-like functions. The presence of EGF and FGF in the culture medium is imperative to inhibit the differentiation of TSCs [76, 77]. Ignatova and colleagues were the first to show the formation of neurospheres with glioma using single-cell cultures in a methylcellulose (MC) matrix in the presence of EGF and FGF [78]. Since then, this technique has been commonly used to study the biology of TSC and the response to treatment [12].

The cells from the neurospheres may be subcultured and the primary clones can generate secondary neurospheres, representing the renewal of the previous population (Fig. 3) [79]. Stem cell renewal has been classified into two types: symmetric, where a division produces two daughter cells or two progenitor cells; or asymmetric, where a division produces one daughter and one progenitor cell [79]. Clonal analysis and serial subcloning assays are critical to definitely identify TSCs.

The neurosphere culture method has also been combined with other cell sorting methods to give a more robust result. Singh and colleagues in their seminal glioma TSC paper sorted the CD133+ cells prior to neurosphere cultures [22]. Similarly, Pavon and colleagues described a more vigorous method to isolate TSC wherein the isolation of neurospheres derived from GBM primary cultures was followed by sorting out the of the CD133+ cells to create further sub-neurospheres [80].

Neurosphere assays, despite their value, also have some limitations. The neurosphere assay is an in vitro phenomenon and does not occur in vivo. Additionally, the act of removing cells from their source and placing them in serum-free cultures may precipitate phenotypic or genetic changes that may not be representative of in vivo behavior. Also, it is important to remember that committed progenitor cells have the ability to produce secondary neurospheres, but they cannot continue to form neurospheres on continued passaging (unlike putative NSC or TSC) [81].

Neurosphere assays are also not useful to calculate stem cell frequency within a sample of tissue as a vast majority somatic cells do not remain viable in the serum-free culture medium and do not form neurospheres [75]. Finally, to establish clonality it is essential for a neurosphere to originate from clone only. However, as shown by Singh and colleagues using time-lapse video microscopy, spheres are highly motile structures with a high incidence of cellular aggregation leading to chimeric neurospheres [82]. In this setting, using a single cell in a miniwell [83], or sparse, widely dispersed cells in MC [84, 85] are viable options.

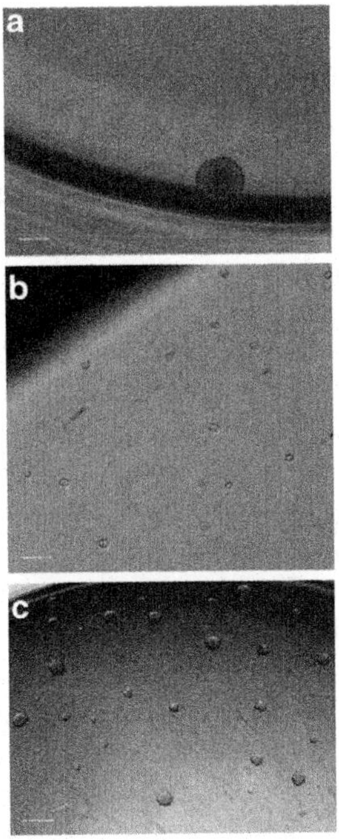

Fig. 3 Individual clonal neurospheres can be subcloned. (**a–c**) Representative examples of SVZ-derived NSCs. Individual clonal spheres generated by limiting dilution from serially passaged adult NSCs (**a**) can be dissociated and single cells replated in the presence of mitogens (**b**), giving rise after 7–10 days to secondary neurospheres (**c**). From 50 up to >200 secondary spheres can be obtained from each individual primary sphere, depending on the number of viable cells plated. The cloning efficiency for adult SVZ-derived NSCs under these culture conditions is in the range of 2–8 %. Bars = 100 μm (**a**), 25 μm (**b**), and 250 μm (**c**). (With permission from Gritti et al. Methods Mol Biol. 2008;438:173–84. Clonal analyses and cryopreservation of neural stem cell cultures)

Orthotopic Implantation

To assess the tumorigenicity of the TSCs, animal orthotopic models are considered the gold standard [86–89]. By allowing growth in in vivo conditions, investigators hope to recapture the cell-to-cell and cell-to-matrix interactions. Intracranial orthotopic implantation of TSCs (in the form of a single cell suspension or neurospheres is mostly conducted in immunocompromised mice [3, 90]). This model is also particularly helpful to study the chemo- and radio-resistant properties of TSC.

Various nude mouse models have been used for implantation assays. The most popular model is the NOD-SCID mouse model, which has a completely knocked out immune system [3]. However this model is susceptible to developing spontaneous lymphomas as early as within a year of life [11]. Other mouse models include IL2 receptor-γ chain deficient, BALB/c-nude and Scid/bg [90, 91].

The most suitable model for in vivo studies is based on the aims of the study and by weighing the advantages against the disadvantages of using that particular model.

To depict self-renewal, the tumor is harvested from one animal and the cells are implanted into another animal. By using fewer and fewer cells in each passage, the enhanced tumorigenicity of TSCs is studied. For neurospheres, Singh and colleagues were the first group to use an in vivo limiting dilution assay to implant fewer and fewer TSCs at each subsequent passage to assess the fewest number of cells required to form a tumor in the animal [22]. As a general principle, compared to the marker-negative population, at least 50- to 100-fold fewer marker-positive cells should be needed to stimulate tumor formation in 50 % of the mice [11]

The mouse implantation model has some limitations as well. First, it is not possible to accurately assess the proportion of TSCs in original tumor mass. This is because the effect of in vivo conditions on the viability and tumorigenic potential is not known [92]. Second, in spite of being an in vivo model, it still has factors such as the extracellular matrix constitution, host immunocompetence, growth factors, and vascularity that may not be representative of the original tumor environment. Finally, the Zebrafish implantation studies for TSC analysis is a relatively newly described functional assay [93, 94]. TSCs are implanted into the peritoneal cavity of the animal in a transparent embryo and the growth and proliferation may be observed directly. Adult transparent zebrafish have also been created for the same purpose [95]. Additional work on developing tissue-specific orthotopic implantation can improve this model further [96].

Conclusion

With increasing evidence pointing towards the importance of TSCs in the biology of CNS tumors, it has become vital to be able to understand the unique biology of these cells and discover potential therapeutic targets. The first step in carrying out such investigation is to reliably isolate and characterize the TSCs. A variety of methods have been described in this regard and researchers need to be aware of the benefits and shortcomings of the methods chosen. In the future, improving technology is anticipated to drive the development of more reliable, accurate, and less labor-intensive assays.

References

1. Stupp R, Mason WP, van den Bent MJ, Weller M, Fisher B, Taphoorn MJB, et al. Radiotherapy plus concomitant and adjuvant temozolomide for glioblastoma. N Engl J Med. 2005;352(10): 987–96.
2. Rich JN, Eyler CE. Cancer stem cells in brain tumor biology. Cold Spring Harb Symp Quant Biol. 2008;73:411–20.

3. Clarke MF, Dick JE, Dirks PB, Eaves CJ, Jamieson CHM, Jones DL, et al. Cancer stem cells: perspectives on current status and future directions—AACR Workshop on cancer stem cells. Cancer Res. 2006;66(19):9339–44.
4. Ehtesham M, Mapara KY, Stevenson CB, Thompson RC. CXCR4 mediates the proliferation of glioblastoma progenitor cells. Cancer Lett. 2009;274(2):305–12.
5. Eyler CE, Rich JN. Survival of the fittest: cancer stem cells in therapeutic resistance and angiogenesis. J Clin Oncol. 2008;26(17):2839–45.
6. Bao S, Wu Q, Sathornsumetee S, Hao Y, Li Z, Hjelmeland AB, et al. Stem cell–like glioma cells promote tumor angiogenesis through vascular endothelial growth factor. Cancer Res. 2006;66(16):7843–8.
7. Rycaj K, Tang DG. Cancer stem cells and radioresistance. Int J Radiat Biol. 2014;90:615–21.
8. Vinogradov S, Wei X. Cancer stem cells and drug resistance: the potential of nanomedicine. Nanomedicine. 2012;7(4):597–615.
9. Grimes C, Margolin DA, Li L. Are cancer stem cells responsible for cancer recurrence? Cell Biol Res Ther [Internet]. 2012 [cited 2014 Jun 10];01(01). Available from: http://www.scitechnol.com /cancer-stem-cells-responsible-for-cancer-recurrence-C6I7.php?article_id=63.
10. Yu Y, Ramena G, Elble RC. The role of cancer stem cells in relapse of solid tumors. Front Biosci (Elite Ed). 2012;4:1528–41.
11. Dobbin ZC, Landen CN. Isolation and characterization of potential cancer stem cells from solid human tumors: potential applications. Curr Protoc Pharmacol. 2013;63:Unit 14.28.
12. Guerrero-Cázares H, Chaichana KL, Quiñones-Hinojosa A. Neurosphere culture and human organotypic model to evaluate brain tumor stem cells. Methods Mol Biol. 2009;568:73–83.
13. Mizrak D, Brittan M, Alison MR. CD133: molecule of the moment. J Pathol. 2008;214(1): 3–9.
14. Corbeil D, Fargeas CA, Huttner WB. Rat prominin, like its mouse and human orthologues, is a pentaspan membrane glycoprotein. Biochem Biophys Res Commun. 2001;285(4):939–44.
15. Weigmann A, Corbeil D, Hellwig A, Huttner WB. Prominin, a novel microvilli-specific polytopic membrane protein of the apical surface of epithelial cells, is targeted to plasmalemmal protrusions of non-epithelial cells. Proc Natl Acad Sci U S A. 1997;94(23):12425–30.
16. Yin AH, Miraglia S, Zanjani ED, Almeida-Porada G, Ogawa M, Leary AG, et al. AC133, a novel marker for human hematopoietic stem and progenitor cells. Blood. 1997;90(12): 5002–12.
17. Miraglia S, Godfrey W, Yin AH, Atkins K, Warnke R, Holden JT, et al. A novel five-transmembrane hematopoietic stem cell antigen: isolation, characterization, and molecular cloning. Blood. 1997;90(12):5013–21.
18. Horn PA, Tesch H, Staib P, Kube D, Diehl V, Voliotis D. Expression of AC133, a novel hematopoietic precursor antigen, on acute myeloid leukemia cells. Blood. 1999;93(4):1435–7.
19. Bühring HJ, Seiffert M, Marxer A, Weiss B, Faul C, Kanz L, et al. AC133 antigen expression is not restricted to acute myeloid leukemia blasts but is also found on acute lymphoid leukemia blasts and on a subset of CD34+ B-cell precursors. Blood. 1999;94(2):832–3.
20. Neuzil J, Stantic M, Zobalova R, Chladova J, Wang X, Prochazka L, et al. Tumour-initiating cells vs. cancer "stem" cells and CD133: what's in the name? Biochem Biophys Res Commun. 2007;355(4):855–9.
21. Singh SK, Clarke ID, Terasaki M, Bonn VE, Hawkins C, Squire J, et al. Identification of a cancer stem cell in human brain tumors. Cancer Res. 2003;63(18):5821–8.
22. Singh SK, Hawkins C, Clarke ID, Squire JA, Bayani J, Hide T, et al. Identification of human brain tumour initiating cells. Nature. 2004;432(7015):396–401.
23. Hemmati HD, Nakano I, Lazareff JA, Masterman-Smith M, Geschwind DH, Bronner-Fraser M, et al. Cancerous stem cells can arise from pediatric brain tumors. Proc Natl Acad Sci U S A. 2003;100(25):15178–83.
24. Dell'Albani P. Stem cell markers in gliomas. Neurochem Res. 2008;33(12):2407–15.
25. Lendahl U, Zimmerman LB, McKay RD. CNS stem cells express a new class of intermediate filament protein. Cell. 1990;60(4):585–95.

26. Hockfield S, McKay RD. Identification of major cell classes in the developing mammalian nervous system. J Neurosci. 1985;5(12):3310–28.
27. Strojnik T, Røsland GV, Sakariassen PO, Kavalar R, Lah T. Neural stem cell markers, nestin and musashi proteins, in the progression of human glioma: correlation of nestin with prognosis of patient survival. Surg Neurol. 2007;68(2):133–43. discussion 143–144.
28. Zhang M, Song T, Yang L, Chen R, Wu L, Yang Z, et al. Nestin and CD133: valuable stem cell-specific markers for determining clinical outcome of glioma patients. J Exp Clin Cancer Res. 2008;27:85.
29. Emmenegger BA, Wechsler-Reya RJ. Stem cells and the origin and propagation of brain tumors. J Child Neurol. 2008;23(10):1172–8.
30. Sotelo C, Alvarado-Mallart RM, Frain M, Vernet M. Molecular plasticity of adult Bergmann fibers is associated with radial migration of grafted Purkinje cells. J Neurosci. 1994; 14(1):124–33.
31. Alder J, Cho NK, Hatten ME. Embryonic precursor cells from the rhombic lip are specified to a cerebellar granule neuron identity. Neuron. 1996;17(3):389–99.
32. Lee A, Kessler JD, Read T-A, Kaiser C, Corbeil D, Huttner WB, et al. Isolation of neural stem cells from the postnatal cerebellum. Nat Neurosci. 2005;8(6):723–9.
33. Pfenninger CV, Roschupkina T, Hertwig F, Kottwitz D, Englund E, Bengzon J, et al. CD133 is not present on neurogenic astrocytes in the adult subventricular zone, but on embryonic neural stem cells, ependymal cells, and glioblastoma cells. Cancer Res. 2007;67(12):5727–36.
34. Eyler CE, Foo W-C, LaFiura KM, McLendon RE, Hjelmeland AB, Rich JN. Brain cancer stem cells display preferential sensitivity to Akt inhibition. Stem Cells. 2008;26(12):3027–36.
35. Pavon LF, Marti LC, Sibov TT, Miyaki LAM, Malheiros SMF, Mamani JB, et al. Isolation, cultivation and characterization of CD133+ stem cells from human glioblastoma. Einstein São Paulo Braz. 2012;10(2):197–202.
36. Shin DH, Xuan S, Kim W-Y, Bae G-U, Kim J-S. CD133 antibody-conjugated immunoliposomes encapsulating gemcitabine for targeting glioblastoma stem cells. J Mater Chem B. 2014;2(24):3771–81.
37. Latt SA, Stetten G, Juergens LA, Willard HF, Scher CD. Recent developments in the detection of deoxyribonucleic acid synthesis by 33258 Hoechst fluorescence. J Histochem Cytochem. 1975;23(7):493–505.
38. Latt SA, Stetten G. Spectral studies on 33258 Hoechst and related bisbenzimidazole dyes useful for fluorescent detection of deoxyribonucleic acid synthesis. J Histochem Cytochem. 1976;24(1):24–33.
39. Goodell MA, Brose K, Paradis G, Conner AS, Mulligan RC. Isolation and functional properties of murine hematopoietic stem cells that are replicating in vivo. J Exp Med. 1996; 183(4):1797–806.
40. Goodell MA, Rosenzweig M, Kim H, Marks DF, DeMaria M, Paradis G, et al. Dye efflux studies suggest that hematopoietic stem cells expressing low or undetectable levels of CD34 antigen exist in multiple species. Nat Med. 1997;3(12):1337–45.
41. Srivastava VK, Nalbantoglu J. Flow cytometric characterization of the DAOY medulloblastoma cell line for the cancer stem-like phenotype. Cytometry A. 2008;73(10):940–8.
42. Shen G, Shen F, Shi Z, Liu W, Hu W, Zheng X, et al. Identification of cancer stem-like cells in the C6 glioma cell line and the limitation of current identification methods. In Vitro Cell Dev Biol Anim. 2008;44(7):280–9.
43. Feuring-Buske M, Hogge DE. Hoechst 33342 efflux identifies a subpopulation of cytogenetically normal CD34(+)CD38(−) progenitor cells from patients with acute myeloid leukemia. Blood. 2001;97(12):3882–9.
44. Kai K, D'Costa S, Yoon B-I, Brody AR, Sills RC, Kim Y. Characterization of side population cells in human malignant mesothelioma cell lines. Lung Cancer Amst Neth. 2010;70(2): 146–51.
45. Holyoake T, Jiang X, Eaves C, Eaves A. Isolation of a highly quiescent subpopulation of primitive leukemic cells in chronic myeloid leukemia. Blood. 1999;94(6):2056–64.

46. She J-J, Zhang P-G, Che X-M, Wang X, Wang Z-M. Side population cells from HXO-Rb44 retinoblastoma cell line have cancer-initiating property. Int J Ophthalmol. 2011;4(5):461–5.
47. Qi W, Zhao C, Zhao L, Liu N, Li X, Yu W, et al. Sorting and identification of side population cells in the human cervical cancer cell line HeLa. Cancer Cell Int. 2014;14(1):3.
48. Xu Y, Xie Y, Wang X, Chen X, Liu Q, Ying M, et al. Identification of cancer stem cells from hepatocellular carcinoma cell lines and their related microRNAs. Oncol Rep. 2013; 30(5):2056–62.
49. Smith PJ, Furon E, Wiltshire M, Campbell L, Feeney GP, Snyder RD, et al. ABCG2-associated resistance to Hoechst 33342 and topotecan in a murine cell model with constitutive expression of side population characteristics. Cytometry A. 2009;75(11):924–33.
50. Zheng X, Shen G, Yang X, Liu W. Most C6 cells are cancer stem cells: evidence from clonal and population analyses. Cancer Res. 2007;67(8):3691–7.
51. Broadley KWR, Hunn MK, Farrand KJ, Price KM, Grasso C, Miller RJ, et al. Side population is not necessary or sufficient for a cancer stem cell phenotype in glioblastoma multiforme. Stem Cells. 2011;29(3):452–61.
52. Marchitti SA, Brocker C, Stagos D, Vasiliou V. Non-P450 aldehyde oxidizing enzymes: the aldehyde dehydrogenase superfamily. Expert Opin Drug Metab Toxicol. 2008;4(6):697–720.
53. Ma I, Allan AL. The role of human aldehyde dehydrogenase in normal and cancer stem cells. Stem Cell Rev. 2011;7(2):292–306.
54. Ginestier C, Wicinski J, Cervera N, Monville F, Finetti P, Bertucci F, et al. Retinoid signaling regulates breast cancer stem cell differentiation. Cell Cycle. 2009;8(20):3297–302.
55. Luo P, Wang A, Payne KJ, Peng H, Wang J, Parrish YK, et al. Intrinsic retinoic acid receptor alpha-cyclin-dependent kinase-activating kinase signaling involves coordination of the restricted proliferation and granulocytic differentiation of human hematopoietic stem cells. Stem Cells. 2007;25(10):2628–37.
56. Duester G, Mic FA, Molotkov A. Cytosolic retinoid dehydrogenases govern ubiquitous metabolism of retinol to retinaldehyde followed by tissue-specific metabolism to retinoic acid. Chem Biol Interact. 2003;143–144:201–10.
57. Hilton J. Role of aldehyde dehydrogenase in cyclophosphamide-resistant L1210 leukemia. Cancer Res. 1984;44(11):5156–60.
58. Kastan MB, Schlaffer E, Russo JE, Colvin OM, Civin CI, Hilton J. Direct demonstration of elevated aldehyde dehydrogenase in human hematopoietic progenitor cells. Blood. 1990; 75(10):1947–50.
59. Jones RJ, Barber JP, Vala MS, Collector MI, Kaufmann SH, Ludeman SM, et al. Assessment of aldehyde dehydrogenase in viable cells. Blood. 1995;85(10):2742–6.
60. Storms RW, Trujillo AP, Springer JB, Shah L, Colvin OM, Ludeman SM, et al. Isolation of primitive human hematopoietic progenitors on the basis of aldehyde dehydrogenase activity. Proc Natl Acad Sci U S A. 1999;96(16):9118–23.
61. Vaidyanathan G, Song H, Affleck D, McDougald DL, Storms RW, Zalutsky MR, et al. Targeting aldehyde dehydrogenase: a potential approach for cell labeling. Nucl Med Biol. 2009;36(8):919–29.
62. Rasper M, Schäfer A, Piontek G, Teufel J, Brockhoff G, Ringel F, et al. Aldehyde dehydrogenase 1 positive glioblastoma cells show brain tumor stem cell capacity. Neuro Oncol. 2010; 12(10):1024–33.
63. Mao P, Joshi K, Li J, Kim S-H, Li P, Santana-Santos L, et al. Mesenchymal glioma stem cells are maintained by activated glycolytic metabolism involving aldehyde dehydrogenase 1A3. Proc Natl Acad Sci U S A. 2013;110(21):8644–9.
64. Zhang W, Yan W, You G, Bao Z, Wang Y, Liu Y, et al. Genome-wide DNA methylation profiling identifies ALDH1A3 promoter methylation as a prognostic predictor in G-CIMP-primary glioblastoma. Cancer Lett. 2013;328(1):120–5.
65. Liu D-Y, Ren C-P, Yuan X-R, Zhang L-H, Liu J, Liu Q, et al. ALDH1 expression is correlated with pathologic grade and poor clinical outcome in patients with astrocytoma. J Clin Neurosci. 2012;19(12):1700–5.

66. Pece S, Tosoni D, Confalonieri S, Mazzarol G, Vecchi M, Ronzoni S, et al. Biological and molecular heterogeneity of breast cancers correlates with their cancer stem cell content. Cell. 2010;140(1):62–73.
67. Dembinski JL, Krauss S. Characterization and functional analysis of a slow cycling stem cell-like subpopulation in pancreas adenocarcinoma. Clin Exp Metastasis. 2009;26(7):611–23.
68. Roesch A, Fukunaga-Kalabis M, Schmidt EC, Zabierowski SE, Brafford PA, Vultur A, et al. A temporarily distinct subpopulation of slow-cycling melanoma cells is required for continuous tumor growth. Cell. 2010;141(4):583–94.
69. Deleyrolle LP, Harding A, Cato K, Siebzehnrubl FA, Rahman M, Azari H, et al. Evidence for label-retaining tumour-initiating cells in human glioblastoma. Brain J Neurol. 2011;134(Pt 5):1331–43.
70. Krafft C, Sobottka SB, Geiger KD, Schackert G, Salzer R. Classification of malignant gliomas by infrared spectroscopic imaging and linear discriminant analysis. Anal Bioanal Chem. 2007; 387(5):1669–77.
71. Krafft C, Thümmler K, Sobottka SB, Schackert G, Salzer R. Classification of malignant gliomas by infrared spectroscopy and linear discriminant analysis. Biopolymers. 2006;82(4): 301–5.
72. Steiner G, Küchler S, Hermann A, Koch E, Salzer R, Schackert G, et al. Rapid and label-free classification of human glioma cells by infrared spectroscopic imaging. Cytometry A. 2008;73A(12):1158–64.
73. Wehbe K, Pineau R, Eimer S, Vital A, Loiseau H, Déléris G. Differentiation between normal and tumor vasculature of animal and human glioma by FTIR imaging. Analyst. 2010; 135(12):3052–9.
74. Uckermann O, Galli R, Anger M, Herold-Mende C, Koch E, Schackert G, et al. Label-free identification of the glioma stem-like cell fraction using Fourier-transform infrared spectroscopy. Int J Radiat Biol. 2014;90:710–7.
75. Reynolds BA, Rietze RL. Neural stem cells and neurospheres: re-evaluating the relationship. Nat Methods. 2005;2(5):333–6.
76. Vescovi AL, Galli R, Reynolds BA. Brain tumour stem cells. Nat Rev Cancer. 2006;6(6): 425–36.
77. Hermann A, Maisel M, Liebau S, Gerlach M, Kleger A, Schwarz J, et al. Mesodermal cell types induce neurogenesis from adult human hippocampal progenitor cells. J Neurochem. 2006;98(2):629–40.
78. Ignatova TN, Kukekov VG, Laywell ED, Suslov ON, Vrionis FD, Steindler DA. Human cortical glial tumors contain neural stem-like cells expressing astroglial and neuronal markers in vitro. Glia. 2002;39(3):193–206.
79. Gritti A, Galli R, Vescovi AL. Clonal analyses and cryopreservation of neural stem cell cultures. Methods Mol Biol. 2008;438:173–84.
80. Pavon LF, Marti LC, Sibov TT, Malheiros SMF, Brandt RA, Cavalheiro S, et al. In vitro analysis of neurospheres derived from glioblastoma primary culture: a novel methodology paradigm. Front Neurol. 2014;4:214.
81. Gritti A, Frölichsthal-Schoeller P, Galli R, Parati EA, Cova L, Pagano SF, et al. Epidermal and fibroblast growth factors behave as mitogenic regulators for a single multipotent stem cell-like population from the subventricular region of the adult mouse forebrain. J Neurosci. 1999; 19(9):3287–97.
82. Singec I, Knoth R, Meyer RP, Maciaczyk J, Volk B, Nikkhah G, et al. Defining the actual sensitivity and specificity of the neurosphere assay in stem cell biology. Nat Methods. 2006;3(10):801–6.
83. Parker MA, Anderson JK, Corliss DA, Abraria VE, Sidman RL, Park KI, et al. Expression profile of an operationally-defined neural stem cell clone. Exp Neurol. 2005;194(2):320–32.
84. Suslov ON, Kukekov VG, Ignatova TN, Steindler DA. Neural stem cell heterogeneity demonstrated by molecular phenotyping of clonal neurospheres. Proc Natl Acad Sci U S A. 2002;99(22):14506–11.

85. Kukekov VG, Laywell ED, Thomas LB, Steindler DA. A nestin-negative precursor cell from the adult mouse brain gives rise to neurons and glia. Glia. 1997;21(4):399–407.
86. Reya T, Morrison SJ, Clarke MF, Weissman IL. Stem cells, cancer, and cancer stem cells. Nature. 2001;414(6859):105–11.
87. Pardal R, Clarke MF, Morrison SJ. Applying the principles of stem-cell biology to cancer. Nat Rev Cancer. 2003;3(12):895–902.
88. Dick JE. Breast cancer stem cells revealed. Proc Natl Acad Sci U S A. 2003;100(7):3547–9.
89. Tysnes BB. Tumor-initiating and -propagating cells: cells that we would like to identify and control. Neoplasia. 2010;12(7):506–15.
90. Galli R, Binda E, Orfanelli U, Cipelletti B, Gritti A, De Vitis S, et al. Isolation and characterization of tumorigenic, stem-like neural precursors from human glioblastoma. Cancer Res. 2004;64(19):7011–21.
91. Choi SA, Lee JY, Phi JH, Wang K-C, Park C-K, Park S-H, et al. Identification of brain tumour initiating cells using the stem cell marker aldehyde dehydrogenase. Eur J Cancer. 2014; 50(1):137–49.
92. Kelly PN, Dakic A, Adams JM, Nutt SL, Strasser A. Tumor growth need not be driven by rare cancer stem cells. Science. 2007;317(5836):337.
93. Dovey MC, Zon LI. Defining cancer stem cells by xenotransplantation in zebrafish. Methods Mol Biol. 2009;568:1–5.
94. Yang X-J, Cui W, Gu A, Xu C, Yu S-C, Li T-T, et al. A novel zebrafish xenotransplantation model for study of glioma stem cell invasion. PLoS One. 2013;8(4):e61801.
95. White RM, Sessa A, Burke C, Bowman T, LeBlanc J, Ceol C, et al. Transparent adult zebrafish as a tool for in vivo transplantation analysis. Cell Stem Cell. 2008;2(2):183–9.
96. Taylor AM, Zon LI. Zebrafish tumor assays: the state of transplantation. Zebrafish. 2009; 6(4):339–46.

The Role of Stem Cells in Pediatric Central Nervous System Malignancies

Branavan Manoranjan, Neha Garg, David Bakhshinyan, and Sheila K. Singh

Abstract Representing the leading cause of childhood cancer mortality, pediatric brain tumors are comprised of diverse histological features, genetic perturbations, cellular populations, treatment protocols, and clinical outcomes. In this chapter we discuss recent and emerging data that implicate cancer stem cells (also known as brain tumor-initiating cells) in initiating and maintaining the growth of a number of pediatric brain tumors including: medulloblastoma, supratentorial primitive

---

B. Manoranjan
McMaster Stem Cell and Cancer Research Institute, McMaster University, Hamilton, ON, Canada L8S 4K1

Michael G. DeGroote School of Medicine, McMaster University, 1200 Main Street West, Hamilton, ON, Canada L8N 3Z5

Department of Biochemistry and Biomedical Sciences, Faculty of Health Sciences, McMaster University, 1200 Main Street West, Hamilton, ON, Canada L8N 3Z5

N. Garg
McMaster Stem Cell and Cancer Research Institute, McMaster University, Hamilton, ON, Canada L8S 4K1

D. Bakhshinyan
McMaster Stem Cell and Cancer Research Institute, McMaster University, Hamilton, ON, Canada L8S 4K1

Department of Biochemistry and Biomedical Sciences, Faculty of Health Sciences, McMaster University, 1200 Main Street West, Hamilton, ON, Canada L8N 3Z5

S.K. Singh, M.D., Ph.D., F.R.C.S(C). (✉)
McMaster Stem Cell and Cancer Research Institute, McMaster University, Hamilton, ON, Canada L8S 4K1

Michael G. DeGroote School of Medicine, McMaster University, 1200 Main Street West, Hamilton, ON, Canada L8N 3Z5

Department of Biochemistry and Biomedical Sciences, Faculty of Health Sciences, McMaster University, 1200 Main Street West, Hamilton, ON, Canada L8N 3Z5

Department of Surgery, Faculty of Health Sciences, McMaster University, 1200 Main Street West, Hamilton, ON, Canada L8N 3Z5
e-mail: ssingh@mcmaster.ca

M. Ehtesham (ed.), *Stem Cell Biology in Neoplasms of the Central Nervous System*, Advances in Experimental Medicine and Biology 853,
DOI 10.1007/978-3-319-16537-0_4

neuroectodermal tumor, atypical teratoid/rhabdoid tumor, ependymoma, low-grade glioma, glioblastoma, diffuse intrinsic pontine glioma, germ cell tumor, and craniopharyngioma. The development of a stem cell framework for the study and treatment of these tumors will enable future clinical approaches to harness the heterogeneous cellular and genomic landscape of these solid tumors as an avenue for developing targeted patient-oriented therapies, thereby improving the overall survivorship for the most lethal childhood cancer.

Keywords Pediatric brain tumor • Cancer stem cell • Brain tumor-initiating cell • Medulloblastoma • Ependymoma • Supratentorial primitive neuroectodermal tumor • Low-grade glioma • Pediatric glioblastoma • Diffuse intrinsic pontine glioma • Atypical teratoid/rhabdoid tumor • Germ cell tumor • Craniopharyngioma

Introduction

Central nervous system (CNS) tumors represent the leading cause of childhood cancer mortality with an incidence of 30 cases per million [1]. Current diagnostic and therapeutic parameters are dependent on clinical history, radiological imaging, and histological confirmation. Although the present era of molecular classification has aided in the recognition of several distinct subgroups in a variety of pediatric brain tumors, the WHO classification of childhood brain tumors remains dependent on tumor location, histopathological features, and immunohistochemical marker expression [1]. These features primarily determine tumor grade, which in turn reflects patient outcome. Prognostically, there have been minimal improvements in the outcome of pediatric brain tumors, but the overall survivorship remains dismal. Consequently, novel biological frameworks must be applied to elucidate mechanisms that may yield high clinical and therapeutic utility.

The cancer stem cell (CSC) hypothesis suggests that a relatively small fraction of tumor cells termed, CSCs, have the ability to proliferate and maintain tumor growth [2]. This is in sharp contrast to all other cells of the bulk tumor, which are characterized by limited proliferative capacity and a more specified lineage potential. More specifically, a CSC maintains two key properties: self-renewal and multilineage differentiation. Self-renewal is defined as the ability of a parental cell to generate an identical daughter cell and a second cell of the same or different phenotype, whereas through the process of differentiation a CSC is able to give rise to the heterogeneous cell lineages that comprise the original tumor [2]. In the recent past, such CSC populations (also termed, tumor-initiating cells, and in the case of brain cancer, brain tumor-initiating cells (BTICs)) have been identified in a number of hematopoietic and solid tumor malignancies based on cell surface markers and stem cell assays, of which the generation of tumors in human–mouse xenograft models has become the gold standard. Moreover, unlike current genomic platforms, the CSC framework takes into account intratumoral heterogeneity by having a developmentally primitive cell at the apex of the hierarchy with a spectrum of more differentiated cells as one goes down this hierarchy [3].

The concept of a CSC suggests that tumors are organized into distinct clonal populations of cells with only the CSC demonstrating the properties of self-renewal and differentiation in vitro and in vivo [2]. Using in vitro assays originally developed to purify neural stem cells (NSCs) [4–6], Singh et al. [7] reported the identification and purification of a cell from primary human medulloblastoma (MB) and glioblastoma (GBM) that had a marked capacity for proliferation, self-renewal, and differentiation. The BTIC represented a minority of tumor cells and was marked by expression of the cell surface marker CD133. Additional in vivo characterization of the BTIC using a human–mouse xenograft assay formally established the identification of CSCs in brain tumors [8]. This work was corroborated in pediatric BTICs, which expressed high levels of NSC genes *CD133*, *Sox2*, *Musashi1*, and *Bmi1*, providing credence to NSC-driven brain tumorigenesis [9]. Furthermore, by taking advantage of unique stem cell properties such as self-renewal, the BTIC model has been clinically validated as a correlative indicator of patient outcome in pediatric brain tumors, suggesting more aggressive tumors to have a higher stem cell self-renewal index. [10] Overall, the BTIC framework of oncogenesis not only takes advantage of developmental genes and pathways implicated in pediatric brain tumorigenesis but also provides an avenue for studying cancer at a cellular level as distinct differences in subsets of tumor cells may not otherwise be appreciated using the current strategy of bulk tumor genomic profiling.

Medulloblastoma

Medulloblastoma (MB) represents the most frequent malignant pediatric brain tumor, comprising 18 % of pediatric intracranial tumors and 350 new diagnoses each year in the United States. The incidence peaks at two timepoints during childhood: 3–4 years and 8–9 years of age [11]. Histologically, MB is classified into several subtypes: classical, desmoplastic/nodular, MB with extensive nodularity, anaplastic, and large cell. Although these subtypes speak to the cellular and morphological heterogeneity of the bulk tumor, their clinical and prognostic utility has remained futile. More recent molecular classifications of MB have re-conceptualized the heterogeneity that exists within these pathological subtypes by identifying multiple distinct molecular subgroups that differ in their demographics, transcriptomes, somatic genetic events, and prognostic outcomes [12–18]. These studies have also given context to the role of key developmental signaling pathways in MB pathogenesis, providing greater support for subgroup-specific BTICs [18–21].

The current consensus for the molecular classification of MB consists of four subgroups, each distinct in terms of prognosis and predicted therapeutic response [18]. Groups 1 and 2 are characterized by upregulation of genes in the Wnt (7–8 % of patients) or Sonic hedgehog (Shh) (28–32 % of patients) pathways, respectively. These two subgroups are associated with improved clinical outcomes, when compared to Groups 3 (26–27 % of patients) and 4 (34–38 % of patients), which are characterized by a greater propensity for metastatic disease and poor clinical outcomes [12–14, 17, 18, 22–24]. Although recent transgenic murine models have identified amplification of *c-myc* signaling to characterize Group 3 MBs [19, 21], both Group 3 and 4 MBs are collectively considered aggressive and refractory to current treatment modalities [25].

Since its initial identification in 1910 by James Homer Wright, MB has been thought to arise from restricted neuronal precursors termed, "neuroblasts" [26]. The concept of a common cell of origin in MB was further supported in 1925 by Percival Bailey and Harvey Cushing in their observation of both glial and neuronal cells, which they proposed to have originated from a primitive embryonic neuroepithelial cell termed, "medulloblast" [27]. Anatomically, the developing cerebellum provides a reservoir of cells susceptible to malignant transformation. Interestingly, pathways implicated in characterizing MB Groups 1 and 2 have also been described in the proliferation, migration, and maturation of cerebellar stem/progenitor cells. In normal cerebellar development, Shh signaling drives the proliferation and migration of a subset of cerebellar precursor cells termed, granule neuron precursors (GNPs). GNPs contain the Shh receptors Ptch and Smo, and thereby respond to a concentration gradient established by the release of Shh ligand from Purkinje cells [28–31]. Genomic alterations in components of the Shh signaling pathway have been identified in up to 25 % of sporadic human MBs and consist of inactivating mutations of *Ptch1* and *Suppressor of fused (Sufu)*, and/or activating mutations of *Smo* [32–36]. The canonical Wnt signaling pathway is responsible for defining the midbrain-hindbrain boundary from which the entire cerebellum develops [37]. More recently, Wnt signaling has been shown to differentially regulate cerebellar NSCs and GNPs [38]. Although Wnt activation in vitro *and* in vivo was shown to promote proliferation of NSCs but not GNPs, the proliferative NSCs did not undergo prolonged expansion or neoplastic growth, suggesting Wnt to function as a regulator of cerebellar stem cell growth and differentiation. Nevertheless, these developmental pathways must be tightly regulated since the cerebellum reaches complete maturation only several months after birth [29], making it a vulnerable target for oncogenic mutations as the developmental phase is prolonged and active beyond in utero.

While the molecular profiling of MB has been credited with providing a developmental approach to studying its pathogenesis, the direct isolation and characterization of subgroup-specific MB BTICs has largely been attributed to several transgenic mouse models. The overlapping expression of genes unique to the human Wnt subgroup in the fetal mouse dorsal brainstem has enabled the discovery of a distinct germinal zone within the hindbrain believed to contain the cell of origin for Wnt-driven MB [20]. Consequently, BTICs may not only contribute to the cellular heterogeneity within MB but may also be responsible for the spatial heterogeneity associated with specific subgroups. Transgenic mice haploinsufficient for *Ptch1* (*Ptch1+/−*), have greatly contributed to elucidating the role of Shh signaling in MB pathogenesis [33]. Through an increase in the proliferative potential of NSCs, the incidence of MB is 15–20 % in these mice [33, 39, 40]. Initial work with this model identified cells resembling GNPs, which retained their proliferative potential suggesting GNPs to promote Shh-dependent MB [33]. More conclusive evidence for the acquisition of a GNP phenotype as being necessary and sufficient for the initiation of Shh-driven MB has been shown with dysregulated Shh-signaling in unipotent

Nestin+GNPs [41], unipotent Math1+ GNPs, or multilineage embryonic NSCs [42, 43]. Although recent evidence has also alluded to a non-cerebellar cell of origin in Group 2 MB [44], similar to that of Wnt-driven MB [20], further characterization of these cell populations is required to truly implicate their involvement in initiating and contributing to MB pathogenesis. While a GNP may very well serve as the cell of origin for Group 2 MB, the identification of tumor propagating cells within this subgroup is required for targeted therapeutic interventions. Interestingly, CD15 has been shown to serve as a putative marker of MB stem cells in the *Ptch1+/−* model [39]. CD15+ cells comprised a small fraction of normal GNPs (as indicated by the co-expression of Math1), and exhibited a higher proliferative capacity, and elevated levels of Shh target genes when compared to CD15- cells. However, unlike CD133+ cells [7, 8], CD15+ cells did not display multilineage differentiation or neurosphere formation when cultured at clonal densities. Consequently, it was believed that these cells marked progenitor populations as opposed to a more primitive stem-like cell. However, upon culturing these cells in serum-free conditions, they were shown to propagate as multipotent MB stem cells, suggesting CD15 to be an additional marker of MB BTICs [40]. Although the evidence for a GNP as the cellular origin for Shh subtype MB is paramount, it remains unclear as to how these cells propagate and transform over the course of tumorigenesis as defined by their regulatory mechanisms and marker expression. The characterization of Group 3 MB BTICs has only recently benefited from discoveries in transgenic mouse models [19, 21]. Using postnatal murine cerebellar stem cells based on the expression of Prominin1 and lacking expression of lineage-specific markers for GNPs, Pei et al. [19] introduced a mutant, stabilized *myc* construct with a dominant negative p53, which induced in vivo MBs distinct from Shh and Wnt murine MBs. In contrast, Kawauchi et al. [21] generated Group 3 MBs in vivo by introducing *myc* ex vivo into *Trp53* null GNPs sorted for the neuronal lineage marker Atoh1 (Math1). Both groups demonstrated the in vivo MBs generated from their transgenic cells to recapitulate many histopathological and genomic features of the human Group 3 MB. Most interestingly, protein and genomic expression profiles of tumors generated by both groups overlapped most with those of NSCs, induced-pluripotent stem cells, and embryonic stem cells. Although Kawauchi et al. [21] had initially injected cells sorted for GNPs negative for stem cell markers such as Prominin1, the resulting tumors had lost Atoh1 expression and instead displayed increased expression of Prominin1 and other stem cell markers. Similarly, Pei et al. [19] observed an increase in markers of undifferentiated cells in the resulting tumors, suggesting Group 3 MB to either arise from cerebellar stem cells or through a process of dedifferentiation in which distinct tumor cells take on a stem-like phenotype. Clinically, the identification of a cerebellar stem cell as the target for initiation and propagation of Group 3 MB is in keeping with the treatment-refractory, metastatic characteristics observed in these patients.

Although Group 4 MB has yet to be described using a transgenic mouse model, stem cell properties and pathways may still be used to further elucidate novel regulatory mechanisms unique to this subgroup. While these tumors share the clinical features of metastatic disease and poor treatment response with Group 3 MBs, their signaling and genomic frameworks are quite distinct [18]. Unlike Group 3 MBs characterized by *myc* signaling, Group 4 MBs are identified by isochromosome 17q and loss of the X chromosome [12–14, 17, 18]. Additional genomics features include the involvement of genes implicated in neuronal development and differentiation. However, the clinical utility and relevance of these genes have yet to be assessed. Given the identification of fate-determination genes in Group 4 MB, the chromatin-modifying Polycomb-group (PcG) gene *Bmi1* has been considered as a novel regulator of Group 4 MB BTICs [45, 46]. Bmi1 functions as a critical regulator of NSC self-renewal through repression of the $p16^{Ink4a}$ and $p19^{Arf}$ senescence pathways [47, 48]. The Bmi1 signaling pathway is also consistently dysregulated or overexpressed in several emerging CSC populations, most recently being cited as a marker of recurrence, poor treatment response, metastatic potential, and death in many cancer models [49, 50]. With respect to MB, *Bmi1* is preferentially expressed in Group 4 tumors [45, 46] and has been shown to recapitulate NSC self-renewal pathways in MB BTICs [45]. The metastatic properties of Group 4 MBs may also be attributed to the interaction of Bmi1 with Twist1 [51]. Twist1, a transcription factor upstream of *Bmi1*, promotes epithelial–mesenchymal transition (EMT) in normal development and metastatic/invasive properties in cancer [51]. Given the preferential expression of both genes in Group 4 MB [45, 46], their interaction may facilitate the invasive and migratory features of Group 4 MB BTICs.

It is apparent that although specific therapeutic targets have yet to be associated with stem cell pathways in MB, several mediators of NSC self-renewal, proliferation, and differentiation continue to demonstrate preferential segregation towards this childhood brain tumor. The continued demonstration of BTICs in transgenic murine models and their evaluation in primary human patient samples will prove to be invaluable in the development of targeted therapies at subgroup-specific BTICs.

Supratentorial Primitive Neuroectodermal Tumor

Supratentorial primitive neuroectodermal tumor (sPNET) accounts for 3–5 % of all pediatric brain tumors and is considered a member of the embryonal family of malignant childhood brain tumors [1, 52]. Although sPNETs resemble the small blue cell histological phenotype attributed to MB, the molecular framework of these tumors is quite distinct. The recent molecular classification of sPNET has categorized this tumor into three distinct subgroups: primitive-neural, oligoneural, and mesenchymal [53]. Clinically, the primitive-neural subgroup represents a younger age of onset (≤4 years), increased metastatic potential within this age group, and the worst overall survivorship among all three subgroups [53]. Of particular interest, developmental genes such as the *Hox* family and pathways including Wnt and Shh are also enriched in the primitive-neural subgroup, establishing the adequate cellular machinery for regulating putative sPNET BTICs. Further evidence for the presence BTICs in maintaining sPNETs is based on the histological heterogeneity observed in these tumors consisting of variable neuronal, ependymal, and glial differentiation [1].

Interestingly, given the strong precedent for the role of BTICs in driving tumorigenesis in sPNETs, a paucity of data remains in the characterization of these cells. Currently, the literature on sPNET BTICs is restricted to one report, in which BTICs were cultured from a human mesenchymal sPNET [54]. The sPNET BTICs from this patient sample were shown to maintain multilineage differentiation into glial, neuronal, and oligodendrocytic lineages along with a sustained self-renewal potential over several in vitro and in vivo passages. CD15+/CD133+ comprised 25–40 % of the bulk tumor population over several passages and demonstrated the greatest in vitro self-renewal capacity. This observation was supported with a reduced overall survival in murine intracranial xenografts of CD15+/CD133+ cells. Therefore, CD15+/CD133+ cells demonstrate a novel cellular target for mesenchymal sPNET BTICs. Future work into the regulation of these cells by pathways enriched in the mesenchymal subgroup such as TGF-β signaling may provide novel small molecules for a tumor that has largely remained refractory to current therapeutic efforts. It also remains to be addressed if the CD15+/CD133+ cells may represent sPNET BTICs irrespective of the molecular subgroup, which may then yield significant clinical utility and improved survivorship for one of the most malignant childhood cancers.

Atypical Teratoid/Rhabdoid Tumor

Atypical Teratoid/Rhabdoid Tumor (AT/RT) is a highly aggressive and malignant intracranial embryonal tumor occurring in children less than 2 years of age [52, 55]. Histologically, these tumors contain a mixture of rhabdoid, primitive neuroepithelial, epithelial, and mesenchymal structures [56]. AT/RT accounts for 2–3 % of all pediatric brain tumors and has a predilection for arising in the posterior fossa [57]. Unlike the heterogeneous histological composition of these tumors, approximately 80 % of AT/RTs contain mutations or deletions in chromosome 22, which account for the inactivation of the *INI1* (*hSNF5/SMARCB1*) gene [58–60]. INI1 functions as a protein component of the ATP-dependent SWI/SNF chromatin-remodeling complex, which regulates genes responsible for proliferation and differentiation [61]. The clinical prognosis is extremely poor with a median survival of 11–17 months, leaving several avenues for future research and targeted therapies at treatment-refractory cell populations [62].

Given the aggressive nature and diverse cell types present in AT/RT, it may be postulated that these tumors contain a BTIC population responsible for maintaining tumor growth, promoting treatment-resistance, and accounting for the distinct cellular architecture [63–65]. Gene expression profiling in CD133+ AT/RT BTICs has shown an increased expression of developmental genes such as *Oct4, Nanog, Sox2, Nestin, Musashi1, and Bmi1*. Interestingly, the identification of drug-resistant/ABC transporter genes including *MDR-1, MRP1,* and *ABCG2* are in keeping with functional data demonstrating CD133+ AT/RT BTICs to be radioresistant [66–68]. Further evidence in support of AT/RT BTICs relates to the overexpression of the chromatin-remodeling Polycomb group complex member, EZH2, which functions to maintain self-renewal, cell growth, proliferation, and radiation-resistance in AT/RT. [69] Although several gene expression studies have been conducted to evaluate the differences between distinct cell populations that may serve as putative AT/RT BTICs, additional mechanistic and in vivo studies may be of greater clinical utility for developing targeted therapies at AT/RT BTICs.

Ependymoma

Ependymoma (EP) is the third most common pediatric brain tumor, representing approximately 9 % of primary brain tumors in children, with an estimated incidence of 200 per year. The median age of diagnosis of pediatric patients is 5 years [11]. Anatomically, EPs occur throughout the CNS, including the supratentorium, posterior fossa, and spinal cord. In the pediatric population, the posterior fossa is the most frequent site of tumorigenesis with 70 % of cases occurring in the fourth ventricle, whereas supratentorial and spinal tumors present more often in adults [70, 71]. According to the WHO, EP may be classified into grades I–III. Although there are distinct anatomical locations for the pathogenesis of EP, histologically, these tumors remain indistinguishable. In keeping with poor histological parameters and a high mortality rate of 45 %, novel genomic markers have been investigated to further delineate genomic subtypes and therapeutic targets [70, 71].

In an attempt to merge genomics with clinical utility, recent molecular analyses of two large independent EP cohorts have revealed the presence of two demographically, transcriptionally, genetically, and clinically distinct groups of pediatric posterior fossa EPs [72]. When compared to Group B patients, Group A patients are younger, associated with laterally located tumors with a balanced genome, and are much more likely to exhibit recurrence, metastasis, and death. Consequently, Group B patients may be treated less aggressively, while novel adjuvant therapies remain critical in the treatment of Group A posterior fossa EPs. Although the most striking candidate markers for distinguishing the two molecular subgroups are *LAMA2* and *NELL2* in Group A and B, respectively, they're functional significance in regulating EP BTICs remains to be investigated.

The first report of EP BTICs was described in an analysis of 100 human EPs, in which the expression of developmental genes was correlated with distinct anatomical origins of tumor formation [73]. Supratentorial, spinal cord, and posterior fossa EPs were found to overexpress members of the *EphB-Ephrin/Notch* pathways, *Hox* gene family, and *AQP1*, respectively. While the pathways and genes associated with supratentorial and spinal cord EPs had been implicated in the regulation of normal NSCs [74] and the anteroposterior patterning of the spinal cord [75], respectively, posterior fossa EPs were continuously found to arise in the SVZ by projecting near the fourth ventricle. Consequently, all three anatomical subgroups demonstrated cell-intrinsic properties in keeping with their anatomically oriented precursor cells.

Through mapping the expression of these genes in the developing mouse, it was found that a distinct type of neural precursor cells termed, radial glial cells, displayed a gene expression profile similar to each of the anatomically distinct human EPs [73]. While this data supported a developmental origin for EP, the generation of a human–mouse EP BTIC xenograft model provided the necessary functional evidence for EP BTICs [73]. Using the NSC markers CD133, Nestin, and RC2 in conjunction with the radial glial cell marker, BLBP, 10,000 CD133+Nestin+ RC2+BLBP+ cells were intracranially injected into the brains of immunocompromised mice. Tumors resembling human EP were identified 4–5 months following transplantation. In contrast, intracranial injections with 2×10 [6] CD133- cells or 2×10 [6] unsorted EP cells did not lead to engraftment or tumor formation even 1 year following transplantation [73]. Consequently, radial glial cells have been considered to function as putative EP BTICs. More recently, tumors resembling human supratentorial EPs have only been capable of developing from the overexpression of EphB2 in mouse embryonic cerebral NSCs [76]. While the genomic anomalies between anatomic subtypes of EP may be distinct, the cellular target continues to function as a BTIC with properties associated with normal radial glial cells. Therefore, continued work in understanding the pathways that promote the differentiation of radial glial cells may be harnessed for therapies targeting the EP BTIC.

Low-Grade Glioma

Pediatric low-grade glioma (LGG) represents the most common pediatric brain tumor, of which the pilocytic astrocytoma (PA) histological subtype accounts for the majority of cases (~20 % of all pediatric brain tumors) [52]. Although these tumors are considered slow-growing and benign, surgically inaccessible midline LGGs remain a therapeutic challenge and account for considerable morbidity and mortality. PAs have classically been described in conjunction with the NF1 (neurofibromatosis) inherited tumor predisposition syndrome [1]. PAs resulting from the mutational inactivation of the *NF1* tumor suppressor gene are primarily located along the optic pathway, while sporadic PAs that do not harbor the *NF1* inactivation predominantly arise in the cerebellum [1]. Recent whole-genome sequencing of PAs has uncovered several recurrent activating mutations in *FGFR1* and *PTPN11* [77, 78]. Although the literature in support of a LGG BTIC is minimal at best, the development of tumors resembling pediatric low-grade optic gliomas from the inactivation of *NF1* in murine third ventricle NSCs has provided a putative cell of origin within the third ventricle for NF1-PAs [79]. Consequently, the application of cell surface markers specific to normal human third ventricle NSCs may further enrich and assist in characterizing putative LGG BTICs. With the discovery of novel LGG driver gene mutations and the presence of LGG BTICs, the future of targeted therapies for inoperable pediatric LGGs remains dependent on the integration of whole-genome sequencing data with cell-intrinsic functional pathways unique to BTICs.

Glioblastoma

Pediatric glioblastoma (GBM) accounts for approximately 3 % of all childhood brain tumors [11, 52]. While pediatric GBMs are histologically identical to adult GBMs, several genomic alterations distinguish these tumors [80, 81]. With the advent of whole-exome sequencing, genomic anomalies unique to pediatric GBM have been identified for the first time [82, 83]. Recurring gain-of-function heterozygous mutations in the *H3F3A* gene, which encodes histone H3.3 have been shown to regulate telomere maintenance and/or heterochromatin stability. Although the comprehensive examination of these mutations in regulating BTICs remains to be investigated, the G34R mutation in *H3F3A* has been shown to promote H3K36me3 enrichment and subsequent activation of transcription factors responsible for NSC proliferation, maintenance, and maturation [84]. In keeping with a primitive transcriptional state, pediatric GBM BTICs have shown an enhanced self-renewal capacity that exceeds septenary spheres with variable expression of putative BTIC markers: CD15, Sox2, Bmi1, Nestin, and Olig2 [85]. The clinical significance of these developmental markers has been established through a *Hox* gene signature that is predictive of temozolomide-resistant pediatric GBM BTICs [86]. Although targeted therapies using cell-intrinsic, treatment-refractory pathways have yet to be pursued with pediatric GBM BTICs, preclinical in vitro and in vivo BTIC models have displayed a reduction in self-renewal capacity and survival advantage in mice, respectively, following treatment with oncolytic viruses [87]. Consequently, pediatric GBMs may provide a novel platform for targeted therapies through the elucidation of epigenetic regulatory mechanisms unique to BTICs that may be amenable to surface marker-based immunotherapies. The integration of diverse research platforms such as cancer genomics, stem cell biology, and immunotherapy may thereby provide a novel paradigm for collaborative research efforts, targeted therapies, and an improvement in the overall survivorship of children diagnosed with pediatric GBM.

Diffuse Intrinsic Pontine Glioma

Diffuse intrinsic pontine glioma (DIPG) is an anatomical variant of high-grade pediatric glioma, which has remained a therapeutic challenge for several decades due to its location in the neurologically delicate brainstem. Brainstem gliomas account for approximately 10–15 % of all pediatric brain tumors, with a median age at presentation of 6–7 years [52, 88]. Clinically, the most common presentation is that of a mass arising in the pons, which is amenable only to radiotherapy. Unfortunately, radiotherapy has shown minimal improvements in mean progression-free survival with an increase to 5.8 months from 5 months for those who do not receive radiotherapy [89]. Overall, 90 % of children succumb to their illness within 2 years of diagnosis, making DIPG one of the leading causes of death in children with brain tumors [88].

Similar to pediatric GBM, whole-genome sequencing has only recently identified novel mutations in the *H3F3A* and *HIST1H3B* genes, which encode histone H3.3 and H3.1, respectively [82, 90]. Histones are basic nuclear proteins responsible for the nucleosome structure of chromosomes. As the nucleosome is formed from DNA being wrapped in repeating units around an octamer consisting of two molecules of each core histone (H2A, H2B, H3, and H4), mutations in histone coding genes provide an epigenetic mechanism of tumor formation. The functional and cellular significance of these mutations in distinct cell populations such as BTICs remains to be established. In keeping with the current era of integrated molecular profiling, DIPGs have recently been classified into two distinct molecular subgroups [91, 92]. Subgroup 1 is associated with activation of the Shh pathway [92] or the presence of mesenchymal/pro-angiogenic markers and the enrichment of developmental genes such as *Sox2, Musashi1,* and *Nestin* [91]. In contrast, subgroup 2 tumors are reflected by *myc* (*N-myc*) activation [92] or the presence of oligodendroglial features with PDGFRA activation [91]. A recent DIPG BTIC human–mouse xenograft model has provided further evidence in establishing a developmental phenotype for subgroup 1 DIPGs [93]. Aside from being regulated by Shh signaling, a regulator of ventral pons precursors, subgroup 1 DIPG BTICs also displayed variable expression of typical BTIC markers (*CD133, Sox2*) along with markers of normal progenitors of the ventral pons (*Nestin, Olig2*). In vivo characterization of the xenograft tumors revealed infiltrative tumors throughout the murine brain involving the cortex, cerebellum, and pons. A second DIPG mouse model in support of a primitive cell of origin for the initiation and maintenance of DIPGs was established using the (RCAS)/tv-a system to overexpress PDGFB in primitive nestin-expressing cells [94]. The malignant transformation of these cells lining the fourth ventricle and aqueduct also lead to the formation of tumors resembling DIPGs. However, unlike human DIPGs, which are thought to arise from the ventral brainstem [93], these transgenic tumors mostly developed from precursor cells near the neonatal dorsal brainstem. The targeted therapy of DIPG BTICs is largely dependent on elucidating the mechanisms that regulate these cells. A recent tissue microarray of human DIPG samples has identified the overexpression of Sox2, Olig2, and Bmi1 in the majority of DIPG samples [95], and thereby provides novel avenues for investigating BTIC mechanisms in a childhood cancer that is only beginning to enter the age of molecular diagnostics, classification, and BTIC-based therapeutics.

Germ Cell Tumor

Intracranial germ cell tumors (iGCTs) represent a rare fraction of pediatric brain tumors, which arise from primordial cells of the developing embryo [96]. Germ cells typically form the reproductive system, but arise in GCTs when they aberrantly migrate and proliferate in sites other than the gonads (i.e., chest, abdomen, and less frequently, the brain) [96]. iGCTs are most commonly found in young people aged 10–20 and account for approximately 2–5 % of all pediatric brain tumors [52, 97].

Gender differences have been reported for both the incidence and localization of iGCTs with a higher incidence in males (3:1) and the localization of tumors in the pineal and suprasellar regions in males and females, respectively [98].

As with many other malignancies, iGCTs possess histopathological subtypes: germinoma, immature/mature teratoma, and non-germinomatous (yolk sac tumor, embryonal carcinoma, and choriocarcinoma) [97, 99, 100]. The primary regimen of treatment includes chemotherapy and/or radiotherapy [101]. Surgery, although not typically recommended due to the inaccessibility of tumors, is preferred in cases of well-encapsulated mature teratomas [96, 102]. Overall prognosis remains relatively poor for non-germinomatous tumors (60 %) when compared to pure germinomas (90 %) [103].

The proposed cell of origin for iGCTs remains controversial as these cells do not maintain a neural lineage. Nevertheless, iGCTs and gonadal GCTs share several molecular features such as chromosomal alterations, mutations in developmental genes, and epigenetic modifications [104]. The expression of several stem cell genes such as *c-kit*, *Oct3/4*, and *Nanog* implicates an embryonic stem cell-like phenotype in these tumors, the hallmark of primordial germ cells [105]. The case for the presence of BTICs in iGCTs has been recently established with the ectopic expression of *Oct4* in NSCs leading to the formation of teratomas in murine xenografts [98, 106]. The identification of elevated Nestin expression, a putative marker for NSCs [107] and BTICs [9], in iGCTs with dissemination and metastatic potential [108] has provided additional evidence for the role of BTICs in driving intracranial germ cell tumorigenesis. Nevertheless, a clear distinction remains to be established between migratory germinal cells or neural lineage-derived BTICs as the cells of origin in iGCTs.

Craniopharyngioma

Craniopharyngioma (CP) is a rare type of low-grade malignancy originating in the sellar and parasellar regions of the brain. 30–50 % of CPs occur in the pediatric population with common symptoms including headaches, visual impairments, growth retardation, and additional symptoms relating to hypothalamic dysfunction [52, 109–111]. Current treatment for CP involves complete tumor resection in cases where the optic nerves or normal functioning of the hypothalamus may not be compromised. However, in cases where complete resection is not possible, surgical resection is complemented by local irradiation [109]. Although current treatment protocols ensure high survival rates (87–95 %), it is common for patients to experience a significant reduction in quality-of-life resulting from surgical complications relating to the optic nerves, pituitary gland, or hypothalamus [112].

CP is a non-glial tumor that originates from the malformation of embryonal tissue [113]. Histopathological features of CP in the pediatric population are in keeping with an adamantinomatous feature with possible cysts and in 70 % of the cases accompanied by stabilizing mutations in *CTNNB1*, which codes for the key downstream effector protein of the canonical Wnt pathway, β-catenin. In contrast, adult cases of CP are in keeping with a squamous-papillary histology [114].

Two possible cellular origins for CP include: the ectopic remnants of Rathke's Pouch (RP) or embryonal epithelial cells of the anterior pituitary gland and infundibulum [115, 116]. A recent transgenic mouse model of adamantinomatous CP with constitutive Wnt pathway activation in progenitor cells of RP has aided in determining a putative TIC [117]. Although all pituitary cells contained the β-catenin mutation, only a small population of cells showed the accumulation of β-catenin in the nucleus and cytosol during the pre-tumoral stages [118]. Further analysis of the β-catenin-enriched fraction of cells revealed the absence of the proliferation marker Ki67 along with the presence of long telomeres—two properties commonly associated with a quiescent state, a feature of stem cells. A subsequent gene expression analysis of the β-catenin-enriched cells showed increased activity of Shh pathway target genes, which are active during cell specification and proliferation of early RP progenitors [119]. Additional immunocytochemical analyses of human CP samples identified an increase in the expression of genes previously implicated in stem cells and BTICs: Sox2, Oct4, KLF4, and Sox9 [118]. Despite the activation of developmental signaling pathways and genes responsible for fate determination, it is still unclear whether RP progenitors may truly function as CP BTICs. The paucity of in vitro and xenograft data from putative human CP BTICs further confounds the identification and characterization of these cells for therapeutic targeting.

Conclusion

The study of pediatric brain tumorigenesis has drastically evolved over the past 100 years, with several key discoveries having been made in only the past decade. With the advent of deep genome sequencing of malignant tissue, the identification of additional molecular classification systems rooted in clinical outcome and risk stratification has begun to emerge. However, the heterogeneous nature of childhood brain tumors remains a burden to be reckoned with as recent reports have shed light on the extent of intratumoral heterogeneity within solid cancers. Given the urgent desire for targeted therapies and the observation of a heterogeneous genomic landscape, other frameworks and model systems should be investigated for exploring the dynamic nature of brain tumors. One such model system is that of the cancer stem cell (CSC) or brain tumor-initiating cell (BTIC). Since rare stem cell populations typically comprise a minority of cells within a heterogeneous tumor and these cells may be underrepresented on bulk tumor analyses, it is possible that very low transcript levels identify critical BTIC regulatory genes when profiling bulk tumors. Consequently, current molecular profiling techniques may not truly account for those genes preferentially expressed within the BTIC population. Moreover, the CSC model provides a framework to study the interplay between BTIC and their tumor niche, offering researchers with multiple perspectives regarding tumor biology and differential gene expression patterns in specific subsets of tumor cells.

Although the CSC model provides several advantages in studying tumor heterogeneity, one must not neglect the limitations accompanied with this framework. These challenges primarily surround our ability to characterize these rare clonal populations of cells, which is particularly true for rare tumors such as MB, pediatric GBM, and DIPG. The current use of cell surface markers to prospectively identify BTICs has proved to be quite controversial. For example, based on differences in cell culture methods CD15 has been shown to identify cells either lacking [39] or displaying [40] multilineage differentiation and neurosphere formation, respectively. CD133 has also recently been identified as a contentious marker for BTICs [120–122]. The original work that prospectively established CD133 as a BTIC marker was restricted to minimally cultured, primary human cells [7, 8]; however, this may not be extended to the long-term culture methodologies applied in several recent papers, in which human BTICs have been passaged as tumor spheres in culture for greater than 3 weeks [123] or more than 20 passages [124]. It is highly possible that these long-term cultured cells have acquired transformation events in vitro that are independent of CD133 expression status or that CD133 protein expression levels no longer correlate with CD133 transcript levels or intracellular receptor activity. Consequently, the utility and readout of cell surface markers in distinguishing BTICs may be highly dependent on cell culture methods. Furthermore, primary human pediatric BTIC cultures are technically challenging, provide limited cell numbers for data acquisition, and require specific infrastructure; therefore, this platform is unlikely to be widely adapted for routine laboratory use at this point, unlike current genomic platforms. However, continued study of larger numbers of human BTIC specimens will eventually elucidate key stem cell signaling pathways and molecular mechanisms of self-renewal that could provide specific targets for tumors that remain refractory to current therapies.

References

1. Louis DN, et al. The 2007 WHO classification of tumours of the central nervous system. Acta Neuropathol. 2007;114:97–109. doi:10.1007/s00401-007-0243-4.
2. Clarke MF, et al. Cancer stem cells–perspectives on current status and future directions: AACR workshop on cancer stem cells. Cancer Res. 2006;66:9339–44. doi:10.1158/0008-5472.CAN-06-3126.
3. Magee JA, Piskounova E, Morrison SJ. Cancer stem cells: impact, heterogeneity, and uncertainty. Cancer Cell. 2012;21:283–96. doi:10.1016/j.ccr.2012.03.003. [pii] S1535-6108(12)00086-4.
4. Reynolds BA, Weiss S. Clonal and population analyses demonstrate that an EGF-responsive mammalian embryonic CNS precursor is a stem cell. Dev Biol. 1996;175:1–13.
5. Reynolds BA, Weiss S. Generation of neurons and astrocytes from isolated cells of the adult mammalian central nervous system. Science. 1992;255:1707–10.
6. Uchida N, et al. Direct isolation of human central nervous system stem cells. Proc Natl Acad Sci U S A. 2000;97:14720–5. doi:10.1073/pnas.97.26.14720.
7. Singh SK, et al. Identification of a cancer stem cell in human brain tumors. Cancer Res. 2003;63:5821–8.
8. Singh SK, et al. Identification of human brain tumour initiating cells. Nature. 2004;432:396–401. doi:10.1038/nature03128. pii: nature03128.
9. Hemmati HD, et al. Cancerous stem cells can arise from pediatric brain tumors. Proc Natl Acad Sci U S A. 2003;100:15178–83. doi:10.1073/pnas.2036535100.
10. Panosyan EH, et al. Clinical outcome in pediatric glial and embryonal brain tumors correlates with in vitro multi-passageable neurosphere formation. Pediatr Blood Cancer. 2010;55:644–51. doi:10.1002/pbc.22627.

11. CBTRUS. Primary brain and central nervous system tumors diagnosed in the United States in 2004–2008. 2012.
12. Cho YJ, et al. Integrative genomic analysis of medulloblastoma identifies a molecular subgroup that drives poor clinical outcome. J Clin Oncol. 2011;29:1424–30. doi:10.1200/JCO.2010.28.5148.
13. Kool M, et al. Integrated genomics identifies five medulloblastoma subtypes with distinct genetic profiles, pathway signatures and clinicopathological features. PLoS ONE. 2008; 3:e3088. doi:10.1371/journal.pone.0003088.t001.
14. Northcott PA, et al. Medulloblastoma comprises four distinct molecular variants. J Clin Oncol. 2011;29:1408–14. doi:10.1200/JCO.2009.27.4324.
15. Pomeroy SL, et al. Prediction of central nervous system embryonal tumour outcome based on gene expression. Nature. 2002;415:436–42. doi:10.1038/415436a.
16. Thompson MC, et al. Genomics identifies medulloblastoma subgroups that are enriched for specific genetic alterations. J Clin Oncol. 2006;24:1924–31. doi:10.1200/JCO.2005.04.4974.
17. Kool M, et al. Molecular subgroups of medulloblastoma: an international meta-analysis of transcriptome, genetic aberrations, and clinical data of WNT, SHH, Group 3, and Group 4 medulloblastomas. Acta Neuropathol. 2012;123:473–84. doi:10.1007/s00401-012-0958-8.
18. Taylor MD, et al. Molecular subgroups of medulloblastoma: the current consensus. Acta Neuropathol. 2012;123:465–72. doi:10.1007/s00401-011-0922-z.
19. Pei Y, et al. An animal model of MYC-driven medulloblastoma. Cancer Cell. 2012;21: 155–67. doi:10.1016/j.ccr.2011.12.021. pii: S1535-6108(11)00483-1.
20. Gibson P et al. Subtypes of medulloblastoma have distinct developmental origins. Nature. 2010;1–5. doi:10.1038/nature09587.
21. Kawauchi D, et al. A mouse model of the most aggressive subgroup of human medulloblastoma. Cancer Cell. 2012;21:168–80. doi:10.1016/j.ccr.2011.12.023. pii: S1535-6108(12)00002-5.
22. Clifford SC, et al. Wnt/Wingless pathway activation and chromosome 6 loss characterize a distinct molecular sub-group of medulloblastomas associated with a favorable prognosis. Cell Cycle. 2006;5:2666–70.
23. Ellison DW, et al. beta-Catenin status predicts a favorable outcome in childhood medulloblastoma: the United Kingdom Children's Cancer Study Group Brain Tumour Committee. J Clin Oncol. 2005;23:7951–7. doi:10.1200/JCO.2005.01.5479.
24. Fattet S, et al. Beta-catenin status in paediatric medulloblastomas: correlation of immunohistochemical expression with mutational status, genetic profiles, and clinical characteristics. J Pathol. 2009;218:86–94. doi:10.1002/path.2514.
25. Ellison DW. Childhood medulloblastoma: novel approaches to the classification of a heterogeneous disease. Acta Neuropathol. 2010;120:305–16. doi:10.1007/s00401-010-0726-6.
26. Wright J. Neurocytoma or neuroblastoma, a kind of tumor not generally recognized. J Exp Med. 1910;12(4):556–61.
27. Bailey P, Cushing H. Medulloblastoma cerebelli: a common type of midcerebellar glioma of childhood. Arch Neurol Psychiatry. 1925;14:192–224. doi:10.1001/archneurpsyc.1925.02200140055002.
28. Fuccillo M, Joyner AL, Fishell G. Morphogen to mitogen: the multiple roles of hedgehog signalling in vertebrate neural development. Nat Rev Neurosci. 2006;7:772–83. doi:10.1038/nrn1990. pii: nrn1990.
29. Wang VY, Zoghbi HY. Genetic regulation of cerebellar development. Nat Rev Neurosci. 2001;2:489–91.
30. Sotelo C. Cellular and genetic regulation of the development of the cerebellar system. Prog Neurobiol. 2004;72:295–339. doi:10.1016/j.pneurobio.2004.03.004. pii: S0301008204000401.
31. Wechsler-Reya RJ, Scott MP. Control of neuronal precursor proliferation in the cerebellum by Sonic Hedgehog. Neuron. 1999;22:103–14.
32. Oliver TG, et al. Loss of patched and disruption of granule cell development in a pre-neoplastic stage of medulloblastoma. Development. 2005;132:2425–39. doi:10.1242/dev.01793.
33. Goodrich LV, Milenkovic L, Higgins KM, Scott MP. Altered neural cell fates and medulloblastoma in mouse patched mutants. Science. 1997;277:1109–13.

34. Hallahan AR. The SmoA1 mouse model reveals that notch signaling is critical for the growth and survival of sonic hedgehog-induced medulloblastomas. Cancer Res. 2004;64:7794–800. doi:10.1158/0008-5472.CAN-04-1813.
35. Raffel C, et al. Sporadic medulloblastomas contain PTCH mutations. Cancer Res. 1997;57: 842–5.
36. Taylor MD, et al. Mutations in SUFU predispose to medulloblastoma. Nat Genet. 2002;31:306–10. doi:10.1038/ng916.
37. Thomas KR, Capecchi MR. Targeted disruption of the murine int-1 proto-oncogene resulting in severe abnormalities in midbrain and cerebellar development. Nature. 1990;346:847–50. doi:10.1038/346847a0.
38. Pei Y et al. WNT signaling increases proliferation and impairs differentiation of stem cells in the developing cerebellum. Development. 2012. doi: 10.1242/dev.050104. pii: dev.050104.
39. Read T-A, et al. Identification of CD15 as a marker for tumor-propagating cells in a mouse model of medulloblastoma. Cancer Cell. 2009;15:135–47. doi:10.1016/j.ccr.2008.12.016.
40. Ward RJ, et al. Multipotent CD15+ cancer stem cells in patched-1-deficient mouse medulloblastoma. Cancer Res. 2009;69:4682–90. doi:10.1158/0008-5472.CAN-09-0342.
41. Li P, et al. A population of Nestin-expressing progenitors in the cerebellum exhibits increased tumorigenicity. Nat Neurosci. 2013;16:1737–44. doi:10.1038/nn.3553. pii: nn.3553.
42. Schüller U, et al. Acquisition of granule neuron precursor identity is a critical determinant of progenitor cell competence to form Shh-induced medulloblastoma. Cancer Cell. 2008;14:123–34. doi:10.1016/j.ccr.2008.07.005.
43. Yang Z-J, et al. Medulloblastoma can be initiated by deletion of patched in lineage-restricted progenitors or stem cells. Cancer Cell. 2008;14:135–45. doi:10.1016/j.ccr.2008.07.003.
44. Grammel D, et al. Sonic hedgehog-associated medulloblastoma arising from the cochlear nuclei of the brainstem. Acta Neuropathol. 2012;123:601–14. doi:10.1007/s00401-012-0961-0.
45. Manoranjan B, et al. FoxG1 interacts with Bmi1 to regulate self-renewal and tumorigenicity of medulloblastoma stem cells. Stem Cells. 2013;31:1266–77. doi:10.1002/stem.1401.
46. Wang X, Venugopal C, Manoranjan B, McFarlane N, O'Farrell E, Nolte S, Gunnarsson T, Hollenberg R, Kwiecien J, Northcott P, Taylor MD, Hawkins C, Singh SK. Sonic hedgehog regulates Bmi1 in human medulloblastoma brain tumor-initiating cells. Oncogene. 2012; 31:187–99.
47. Bruggeman SWM, et al. Bmi1 controls tumor development in an Ink4a/Arf-independent manner in a mouse model for glioma. Cancer Cell. 2007;12:328–41. doi:10.1016/j.ccr.2007.08.032.
48. Leung C, Lingbeek M, Shakhova O, Liu J, Tanger E, Saremaslani P, Van Lohuizen M, Marino S. Bmi1 is essential for cerebellar development and is overexpressed in human medulloblastomas. Nature. 2004;428:337–41.
49. Glinsky GV, Berezovska O, Glinskii AB. Microarray analysis identifies a death-from-cancer signature predicting therapy failure in patients with multiple types of cancer. J Clin Invest. 2005;115:1503–21. doi:10.1172/JCI23412DS1.
50. Zakrzewska M, et al. Polycomb genes expression as a predictor of poor clinical outcome in children with medulloblastoma. Childs Nerv Syst. 2010;27:79–86. doi:10.1007/s00381-010-1260-5.
51. Yang M-H, et al. Bmi1 is essential in Twist1-induced epithelial–mesenchymal transition. Nat Cell Biol. 2010;12:982–92. doi:10.1038/ncb2099.
52. Dolecek TA, Propp JM, Stroup NE, Kruchko C. CBTRUS statistical report: primary brain and central nervous system tumors diagnosed in the United States in 2005–2009. Neuro Oncol. 2012;14 Suppl 5:v1–49. doi:10.1093/neuonc/nos218. pii: nos218.
53. Picard D, et al. Markers of survival and metastatic potential in childhood CNS primitive neuro-ectodermal brain tumours: an integrative genomic analysis. Lancet Oncol. 2012;13:838–48. doi:10.1016/S1470-2045(12)70257-7. pii: S1470-2045(12)70257-7.
54. Liu Z et al. A patient tumor-derived orthotopic xenograft mouse model replicating the group 3 supratentorial primitive neuroectodermal tumor in children. Neuro Oncol. 2014. doi: 10.1093/neuonc/not244, pii: not244.

55. Cocce MC, Lubieniecki F, Kordes U, Alderete D, Gallego MS. A complex karyotype in an atypical teratoid/rhabdoid tumor: case report and review of the literature. J Neuro-Oncol. 2011;104:375–80. doi:10.1007/s11060-010-0478-0.
56. Avci Z, Kaya IS, Dogukan AS, Aydin O, Ismailoglu O. Pathology teach and tell: central nervous system atypical teratoid/rhabdoid tumor. Pediatr Pathol Mol Med. 2003;22:443–7.
57. Rorke LB, Packer RJ, Biegel JA. Central nervous system atypical teratoid/rhabdoid tumors of infancy and childhood: definition of an entity. J Neurosurg. 1996;85:56–65. doi:10.3171/jns.1996.85.1.0056.
58. Versteege I, et al. Truncating mutations of hSNF5/INI1 in aggressive paediatric cancer. Nature. 1998;394:203–6. doi:10.1038/28212.
59. Pfister SM, et al. Molecular diagnostics of CNS embryonal tumors. Acta Neuropathol. 2010;120:553–66. doi:10.1007/s00401-010-0751-5.
60. Ichimura K, Nishikawa R, Matsutani M. Molecular markers in pediatric neuro-oncology. Neuro Oncol 2012;14(Suppl 4): iv90–9, doi:10.1093/neuonc/nos204.
61. Wilson BG, Roberts CW. SWI/SNF nucleosome remodellers and cancer. Nat Rev Cancer. 2011;11:481–92. doi:10.1038/nrc3068.
62. Nicolaides T, et al. High-dose chemotherapy and autologous stem cell rescue for atypical teratoid/rhabdoid tumor of the central nervous system. J Neuro-Oncol. 2010;98:117–23. doi:10.1007/s11060-009-0071-6.
63. Yachnis AT, Neubauer D, Muir D. Characterization of a primary central nervous system atypical teratoid/rhabdoid tumor and derivative cell line: immunophenotype and neoplastic properties. J Neuropathol Exp Neurol. 1998;57:961–71.
64. Biegel JA. Molecular genetics of atypical teratoid/rhabdoid tumor. Neurosurg Focus. 2006;20:E11.
65. Parham DM, Weeks DA, Beckwith JB. The clinicopathologic spectrum of putative extrarenal rhabdoid tumors. An analysis of 42 cases studied with immunohistochemistry or electron microscopy. Am J Surg Pathol. 1994;18:1010–29.
66. Chiou SH, et al. Identification of CD133-positive radioresistant cells in atypical teratoid/rhabdoid tumor. PLoS ONE. 2008;3:e2090. doi:10.1371/journal.pone.0002090.
67. Kao CL, et al. Resveratrol-induced apoptosis and increased radiosensitivity in CD133-positive cells derived from atypical teratoid/rhabdoid tumor. Int J Radiat Oncol Biol Phys. 2009;74:219–28. doi:10.1016/j.ijrobp.2008.12.035.
68. Ma HI, et al. Differential expression profiling between atypical teratoid/rhabdoid and medulloblastoma tumor in vitro and in vivo using microarray analysis. Childs Nerv Syst. 2010;26:293–303. doi:10.1007/s00381-009-1016-2.
69. Alimova I, et al. Inhibition of EZH2 suppresses self-renewal and induces radiation sensitivity in atypical rhabdoid teratoid tumor cells. Neuro-Oncology. 2013;15:149–60. doi:10.1093/neuonc/nos285.
70. Mack SC, Taylor MD. The genetic and epigenetic basis of ependymoma. Childs Nerv Syst. 2009;25:1195–201. doi:10.1007/s00381-009-0928-1.
71. Kilday JP, et al. Pediatric ependymoma: biological perspectives. Mol Cancer Res. 2009;7:765–86. doi:10.1158/1541-7786.MCR-08-0584. pii: 1541-7786.MCR-08-0584.
72. Witt H, et al. Delineation of two clinically and molecularly distinct subgroups of posterior fossa ependymoma. Cancer Cell. 2011;20:143–57. doi:10.1016/j.ccr.2011.07.007. pii: S1535-6108(11)00262-5.
73. Taylor MD, et al. Radial glia cells are candidate stem cells of ependymoma. Cancer Cell. 2005;8:323–35. doi:10.1016/j.ccr.2005.09.001. pii: S1535-6108(05)00270-9.
74. Conover JC, et al. Disruption of Eph/ephrin signaling affects migration and proliferation in the adult subventricular zone. Nat Neurosci. 2000;3:1091–7. doi:10.1038/80606.
75. Dasen JS, Liu JP, Jessell TM. Motor neuron columnar fate imposed by sequential phases of Hox-c activity. Nature. 2003;425:926–33. doi:10.1038/nature02051. pii: nature02051.
76. Johnson RA, et al. Cross-species genomics matches driver mutations and cell compartments to model ependymoma. Nature. 2010;466:632–6. pii: nature 09173.
77. Jones DT, et al. Recurrent somatic alterations of FGFR1 and NTRK2 in pilocytic astrocytoma. Nat Genet. 2013;45:927–32. doi:10.1038/ng.2682. pii: ng.2682.

78. Zhang J, et al. Whole-genome sequencing identifies genetic alterations in pediatric low-grade gliomas. Nat Genet. 2013;45:602–12. doi:10.1038/ng.2611. pii: ng.2611.
79. da Lee Y, Gianino SM, Gutmann DH. Innate neural stem cell heterogeneity determines the patterning of glioma formation in children. Cancer Cell. 2012;22:131–8. doi:10.1016/j.ccr.2012.05.036. pii: S1535-6108(12)00251-6.
80. Paugh BS, et al. Integrated molecular genetic profiling of pediatric high-grade gliomas reveals key differences with the adult disease. J Clin Oncol. 2010;28:3061–8. doi:10.1200/JCO.2009.26.7252. pii: JCO.2009.26.7252.
81. Pollack IF, et al. Age and TP53 mutation frequency in childhood malignant gliomas: results in a multi-institutional cohort. Cancer Res. 2001;61:7404–7.
82. Wu G, et al. Somatic histone H3 alterations in pediatric diffuse intrinsic pontine gliomas and non-brainstem glioblastomas. Nat Genet. 2012;44:251–3. doi:10.1038/ng.1102. pii: ng.1102.
83. Schwartzentruber J, et al. Driver mutations in histone H3.3 and chromatin remodelling genes in paediatric glioblastoma. Nature. 2012;482:226–31. pii: nature 10833.
84. Bjerke L et al. Histone H3.3 mutations drive pediatric glioblastoma through upregulation of MYCN. Cancer Discov. 2013. doi: 10.1158/2159-8290.CD-12-0426. pii: 2159-8290.CD-12-0426.
85. Thirant C, et al. Clinical relevance of tumor cells with stem-like properties in pediatric brain tumors. PLoS ONE. 2011;6:e16375. doi:10.1371/journal.pone.0016375.
86. Gaspar N, et al. MGMT-independent temozolomide resistance in pediatric glioblastoma cells associated with a PI3-kinase-mediated HOX/stem cell gene signature. Cancer Res. 2010;70:9243–52. doi:10.1158/0008-5472.CAN-10-1250. pii: 0008-5472.CAN-10-1250.
87. Liu Z, et al. Intravenous injection of oncolytic picornavirus SVV-001 prolongs animal survival in a panel of primary tumor-based orthotopic xenograft mouse models of pediatric glioma. Neuro Oncol. 2013;15:1173–85. doi:10.1093/neuonc/not065. pii: not065.
88. Hargrave D, Bartels U, Bouffet E. Diffuse brainstem glioma in children: critical review of clinical trials. Lancet Oncol. 2006;7:241–8. doi:10.1016/S1470-2045(06)70615-5. pii: S1470-2045(06)70615-5.
89. Schroeder KM, Hoeman CM, Becher OJ. Children are not just little adults: recent advances in understanding of diffuse intrinsic pontine glioma biology. Pediatr Res. 2013. doi: 10.1038/pr.2013.194. pii: pr2013194.
90. Khuong-Quang DA, et al. K27M mutation in histone H3.3 defines clinically and biologically distinct subgroups of pediatric diffuse intrinsic pontine gliomas. Acta Neuropathol. 2012;124:439–47. doi:10.1007/s00401-012-0998-0.
91. Puget S, et al. Mesenchymal transition and PDGFRA amplification/mutation are key distinct oncogenic events in pediatric diffuse intrinsic pontine gliomas. PLoS ONE. 2012;7:e30313. doi:10.1371/journal.pone.0030313. pii: PONE-D-11-11908.
92. Saratsis AM, et al. Comparative multidimensional molecular analyses of pediatric diffuse intrinsic pontine glioma reveals distinct molecular subtypes. Acta Neuropathol. 2013. doi:10.1007/s00401-013-1218-2.
93. Monje M, et al. Hedgehog-responsive candidate cell of origin for diffuse intrinsic pontine glioma. Proc Natl Acad Sci U S A. 2011;108:4453–8. doi:10.1073/pnas.1101657108. pii: 1101657108.
94. Becher OJ, et al. Preclinical evaluation of radiation and perifosine in a genetically and histologically accurate model of brainstem glioma. Cancer Res. 2010;70:2548–57. doi:10.1158/0008-5472.CAN-09-2503. pii: 0008-5472.CAN-09-2503.
95. Ballester LY, et al. Morphologic characteristics and immunohistochemical profile of diffuse intrinsic pontine gliomas. Am J Surg Pathol. 2013;37:1357–64. doi:10.1097/PAS.0b013e318294e817. pii: 00000478-201309000-00008.
96. Thakkar JP, Chew L, Villano JL. Primary CNS germ cell tumors: current epidemiology and update on treatment. Med Oncol. 2013;30:496. doi:10.1007/s12032-013-0496-9.
97. Khatua S, Sadighi ZS, Pearlman ML, Bochare S, Vats TS. Brain tumors in children–current therapies and newer directions. Indian J Pediatr. 2012;79:922–7. doi:10.1007/s12098-012-0689-9.

98. Tan C, Scotting PJ. Stem cell research points the way to the cell of origin for intracranial germ cell tumours. J Pathol. 2013;229:4–11. doi:10.1002/path.4098.
99. Herrmann HD, Westphal M, Winkler K, Laas RW, Schulte FJ. Treatment of nongerminomatous germ-cell tumors of the pineal region. Neurosurgery. 1994;34:524–9. discussion 529.
100. Calaminus G, et al. Intracranial germ cell tumors: a comprehensive update of the European data. Neuropediatrics. 1994;25:26–32. doi:10.1055/s-2008-1071577.
101. Bamberg M, et al. Radiation therapy for intracranial germinoma: results of the German cooperative prospective trials MAKEI 83/86/89. J Clin Oncol. 1999;17:2585–92.
102. Schild SE, et al. Histologically confirmed pineal tumors and other germ cell tumors of the brain. Cancer. 1996;78:2564–71.
103. Kyritsis AP. Management of primary intracranial germ cell tumors. J Neuro-Oncol. 2010;96:143–9. doi:10.1007/s11060-009-9951-z.
104. Sakuma Y, et al. c-kit gene mutations in intracranial germinomas. Cancer Sci. 2004;95: 716–20.
105. Hoei-Hansen CE, et al. New evidence for the origin of intracranial germ cell tumours from primordial germ cells: expression of pluripotency and cell differentiation markers. J Pathol. 2006;209:25–33. doi:10.1002/path.1948.
106. Scotting PJ. Are cranial germ cell tumours really tumours of germ cells? Neuropathol Appl Neurobiol. 2006;32:569–74. doi:10.1111/j.1365-2990.2006.00797.x.
107. Wiese C, et al. Nestin expression–a property of multi-lineage progenitor cells? Cell Mol Life Sci. 2004;61:2510–22. doi:10.1007/s00018-004-4144-6.
108. Sakurada K, et al. Nestin expression in central nervous system germ cell tumors. Neurosurg Rev. 2008;31:173–6. doi:10.1007/s10143-007-0115-3. discussion 176–177.
109. Muller HL. Childhood craniopharyngioma. Pituitary. 2013;16:56–67. doi:10.1007/s11102-012-0401-0.
110. Wisoff JH. Craniopharyngioma. J Neurosurg Pediatr. 2008;1:124–5. doi:10.3171/PED/2008/1/2/124. discussion 125.
111. Elliott RE, Wisoff JH. Surgical management of giant pediatric craniopharyngiomas. J Neurosurg Pediatr. 2010;6:403–16. doi:10.3171/2010.8.PEDS09385.
112. Karavitaki N, Cudlip S, Adams CB, Wass JA. Craniopharyngiomas. Endocr Rev. 2006;27:371–97. doi:10.1210/er.2006-0002. pii: er.2006-0002.
113. Garre ML, Cama A. Craniopharyngioma: modern concepts in pathogenesis and treatment. Curr Opin Pediatr. 2007;19:471–9. doi:10.1097/MOP.0b013e3282495a22. pii: 00008480-200708000-00016.
114. Holsken A, Buchfelder M, Fahlbusch R, Blumcke I, Buslei R. Tumour cell migration in adamantinomatous craniopharyngiomas is promoted by activated Wnt-signalling. Acta Neuropathol. 2010;119:631–9. doi:10.1007/s00401-010-0642-9.
115. Nielsen EH, et al. Incidence of craniopharyngioma in Denmark (n = 189) and estimated world incidence of craniopharyngioma in children and adults. J Neurooncol. 2011;104:755–63. doi:10.1007/s11060-011-0540-6.
116. Bunin GR, et al. The descriptive epidemiology of craniopharyngioma. J Neurosurg. 1998;89:547–51. doi:10.3171/jns.1998.89.4.0547.
117. Gaston-Massuet C, et al. Increased wingless (Wnt) signaling in pituitary progenitor/stem cells gives rise to pituitary tumors in mice and humans. Proc Natl Acad Sci U S A. 2011;108:11482–7. doi:10.1073/pnas.1101553108. pii: 1101553108.
118. Andoniadou CL, et al. Identification of novel pathways involved in the pathogenesis of human adamantinomatous craniopharyngioma. Acta Neuropathol. 2012;124:259–71. doi:10.1007/s00401-012-0957-9.
119. Zhu X, Gleiberman AS, Rosenfeld MG. Molecular physiology of pituitary development: signaling and transcriptional networks. Physiol Rev. 2007;87:933–63. doi:10.1152/physrev.00006.2006. pii: 87/3/933.
120. Beier D, et al. CD133(+) and CD133(-) glioblastoma-derived cancer stem cells show differential growth characteristics and molecular profiles. Cancer Res. 2007;67:4010–5. doi:10.1158/0008-5472.CAN-06-4180. pii: 67/9/4010.

121. Clement V, Dutoit V, Marino D, Dietrich PY, Radovanovic I. Limits of CD133 as a marker of glioma self-renewing cells. Int J Cancer. 2009;125:244–8. doi:10.1002/ijc.24352.
122. Sun Y, et al. CD133 (Prominin) negative human neural stem cells are clonogenic and tripotent. PLoS ONE. 2009;4:e5498. doi:10.1371/journal.pone.0005498.
123. Abdouh M, et al. BMI1 sustains human glioblastoma multiforme stem cell renewal. J Neurosci. 2009;29:8884–96. doi:10.1523/JNEUROSCI.0968-09.2009. pii: 29/28/8884.
124. Chen R, et al. A hierarchy of self-renewing tumor-initiating cell types in glioblastoma. Cancer Cell. 2010;17:362–75. doi:10.1016/j.ccr.2009.12.049. pii: S1535-6108(10)00065-6.

Laboratory Models for Central Nervous System Tumor Stem Cell Research

Imad Saeed Khan and Moneeb Ehtesham

Abstract Central nervous system (CNS) tumors are complex organ systems comprising of a neoplastic component with associated vasculature, inflammatory cells, and reactive cellular and extracellular components. Research has identified a subset of cells in CNS tumors that portray defining properties of neural stem cells, namely, that of self-renewal and multi-potency. Growing evidence suggests that these tumor stem cells (TSC) play an important role in the maintenance and growth of the tumor. Furthermore, these cells have also been shown to be refractory to conventional therapy and may be crucial for tumor recurrence and metastasis. Current investigations are focusing on isolating these TSC from CNS tumors to investigate their unique biological processes. This understanding will help identify and develop more effective and comprehensive treatment strategies. This chapter provides an overview of some of the most commonly used laboratory models for CNSTSC research.

Keywords Glioma stem cells • Tumor stem cells • Cancer stem cells • Laboratory models • Neurosphere culture • Matrigel-based assays • Orthotopic culture • Mathematical models • Animal models

Introduction

As evidenced by the name Glioblastoma Multiforme, investigators have long recognized the morphological variation depicted by brain tumors. Brain tumors have been described as complex organ systems comprising of a neoplastic component with associated vasculature, inflammatory cells, and reactive cellular and extracellular components. Tumor genetic analyses have also demonstrated regional

I.S. Khan, M.D.
Section of Neurosurgery, Geisel School of Medicine at Dartmouth, Lebanon, NH, USA

M. Ehtesham, M.D. (✉)
Department of Neurological Surgery, Vanderbilt University Medical Center, Nashville, TN, USA
e-mail: moneeb.ehtesham@Vanderbilt.Edu

M. Ehtesham (ed.), *Stem Cell Biology in Neoplasms of the Central Nervous System*, Advances in Experimental Medicine and Biology 853,
DOI 10.1007/978-3-319-16537-0_5

variation in gene expression and chromosomal aberrations. Using differentiation markers, multiple states and variations have also been shown in brain tumors [1].

There has been increasing interest in the role of tumor stem cells (TSC) in the pathogenesis of CNS tumors. This has given rise to the cancer stem cell tumor model that is predicated upon the presence of a small subset of cancer cells with the exclusive ability to divide and expand the TSC pool, and also to give rise of heterogeneous non-tumorigenic cancer cell lineages that constitute the bulk of the tumor [2].

TSC (also known as cancer stem cells or tumor-initiating cells) were first isolated from blood cancers. A small fraction of acute myeloid leukemia (AML) cells were shown to be capable of initiating and sustaining clonogenic growth and inducing leukemia in nonobese diabetic (NOD)/severe combined immunodeficient (SCID) mice [3, 4]. Of note, these leukemic subclones shared the same cell surface markers (CD43+, CD38−) as hematopoietic stem cells, while the progeny of these clones, the blast cells, often expressed more differentiated lymphoid or myeloid lineage markers. More recently, cancer stem cells have also been isolated from gliomas, gliosarcomas, medulloblastomas, and ependymomas [5–12].

Importantly, increasing evidence suggests that while current cytotoxic therapeutics may kill the bulk of cancer cells, they are often not able to eliminate the critical TSC, which are protected by specific innate resistance mechanisms [13, 14]. The surviving TSC can then account for tumor recurrence or metastasis [15, 16]. The recurrent tumors are resistant to previously used therapeutic modalities and lead to a worsened prognosis for the patient. These considerations may explain the lack of success with current treatments for gliomas and stress the importance of studying the biological processes of TSC to identify potential therapeutic targets.

Unfortunately, no particular markers or gene expression signatures associated with TSC alone have been identified till now. For example, CD133 was previously thought to be a robust CNSTSC marker, but recent work has shown that the marker does not consistently distinguish tumorigenic from nontumorigenic cells [17–19]. While sorting techniques may be used to aid in the isolation and identification of cells, the stemness of a cell may only be confirmed with functional assays. These assays must be able to depict the TSC properties of self-renewal and lineage capacity. Two major assays are used for the enrichment of TSC: neurosphere cultures with multiple passes and animal propagation studies. We provide an overview of these assays, along with some of the other laboratory models used to study the properties of the CNS TSC.

Neurosphere Cultures

Neurospheres characterize three dimensional in vitro spheroid cell clusters that form when mitotic cells of the mammalian CNS are placed in a serum-free medium on a non-adhesive substrate. This assay was first developed for neural stem cells but has been used for TSC isolation and research as well. These spheres generally have the least differentiated cellular populations located on the surface, with cells expressing differentiation in the interior (Fig. 1) [1].

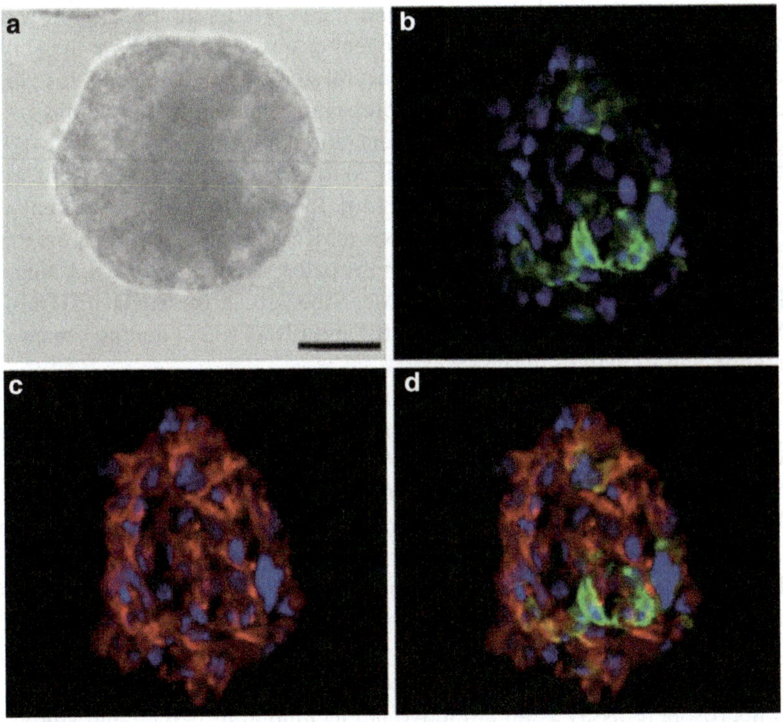

Fig. 1 GBM-derived neurosphere. (**a**) In bright field prior to the cryosectioning protocol described; (**b–d**) immunostained against GFAP (*green*) and Nestin (*red*). Dapi was used as a nuclear marker. The finding of more differentiated cells in the core of the neurosphere showing the cell heterogeneity within the neurospheres (with permission from Guerrero-Cázares H et al. Methods Mol Biol. 2009;568:73–83. Neurosphere culture and human organotypic model to evaluate brain tumor stem cells)

The neurosphere assay relies on a culture system that selectively supports the survival and proliferation of stem and progenitor cells that respond to epidermal growth factor (EGF), basic fibroblast growth factor-2 (FGF-2), or both, forming clonal aggregates called primary neurospheres [20]. EGF in mice has been shown to increase proliferation and survival of precursor cells in the subventricular zone (SVC), and also enhance the generation of astrocytes [21, 22]. FGF also increases neurogenesis, but is additionally protective against injury-induced degeneration [23–25]. Nakano and colleagues have studied the relative roles of stimulation of FGF and EGF receptors on self-renewal of neural stem cells and found that FGF induces a greater degree of self-renewal than EGF family members in embryonic cortical NSC [26]. Other growth factors that simulate the receptors of EGF and FGF also induce the production of neurospheres [27, 28]. Conversely, the removal of these mitogens has been shown to induce differentiation [29, 30].

Upon subculturing the primary neurosphere, renewal of the previous cells may be demonstrated by the production of secondary neurospheres [31]. Certainly, under these conditions, growth factor-responsive cells can be long-term passaged, maintaining stem cell characteristics of multi-potency and stable proliferation [32] However it is important to consider that both TSC and progenitor non-stem cells have the ability to proliferate with each passage. To recognize neural stem cells it is imperative to display their characteristic nature of self-renewal: propagation in long-term cultures (at least five passes) and multipotency through the generation of its progeny [20].

The neurosphere assay can also be modified to investigate the function of various genes by introducing transient transfection [26]. Investigators have used small interfering RNA (siRNA) and small hairpin RNA (shRNA) to interfere with the expression of various genes to assess their roles in the pathogenesis of tumors. Using these techniques a growing body of work has described roles for HEDGEHOG-GLI1 [33], SMC1A [34], ASPP [35], hTERT [36], and FRAT1 [37] among others in the proliferation and growth of gliomas.

The effect of a particular treatment on neurospheres may be analyzed via a variety of methods. The number of clonal spheres a week after treatment corresponds to the proliferative capacity of neurosphere. Alternatively, the number of proliferating cells can also be assessed with flow cytometry based BrdU incorporation labeling [38]. We have also described quantitation of neural stem cell and TSC viability using a colorimetric assay for mitochondrial dehydrogenase activity [38, 39]. The size of spheres can also be quantified (proportional to their diameters) as a measurement of the effect of the treatment on asymmetric self-renewal total neural progenitor proliferation. Moreover, by conducting immunocytochemistry, the effect of on differentiation capacity of treated progenitors may be investigated [26].

The neurosphere assay is an attractive lab technique to isolate and study brain TSC because it includes the functional assay as the initial step [1]. A variant of the neurosphere culture, called the Cambridge Protocol, combines neurosphere and monolayer culture techniques in a bid to improve the efficiency with which cells can be derived from tumor samples under serum-free conditions [40]. The mainstay of neurosphere assay is the assumption that the neuroshere cells are clonal and there is no contribution from any other cell lines [41, 42]. However, generating single cells for neuropshere assays is very challenging and incompletely disintegrated cells may lead to sphere formation with chimerism. Additionally not all cells that form neurospheres are stem cells, as committed progenitor cells also have the same ability. Because clonality of spheres is of immense importance to stem cell research, the clonal relationships need to be confirmed with additional methods such as retroviral marking [41], using a single cell in a miniwell [42], or sparse, widely dispersed cells in methylcellulose [43, 44]. Once clonality has been established, stem cell lines can be effectively expanded and be cryopreserved. Of note, repeated cycles of freezing and thawing do not seem to affect the neural stem cell functional properties [1].

For proliferative studies, the diameter of neurospheres may also not be a very reliable marker [45]. While some of the earlier studies have shown the formation of large neurospheres within 7 days after culture, recent work suggests that it is virtually impossible to produce a large neurosphere in 7 days only [46, 47]; under stringent conditions large neurospheres are only detectable 2–3 weeks after culture. Certainly, a closer look has shown that spheres sometimes integrate free cells, or fuse with other spheres in the clonal medium, leading to chimeric spheres with rapid growth [45] (Fig. 2).

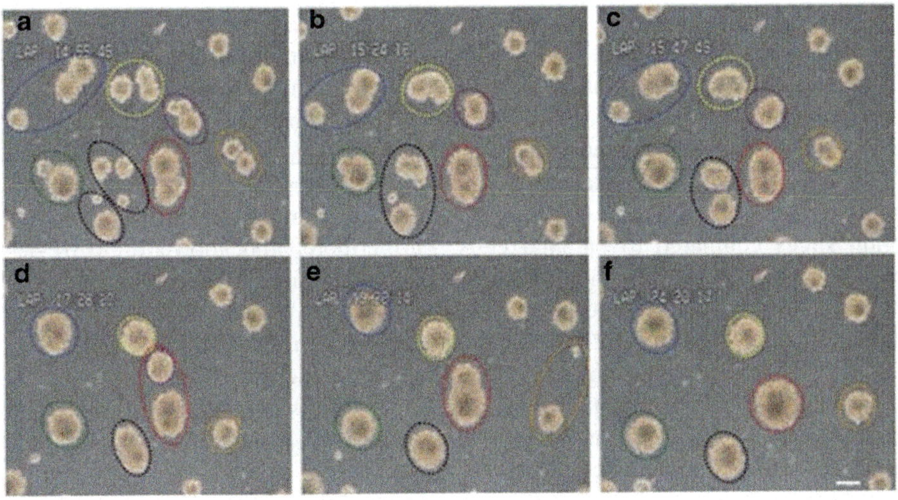

Fig. 2 Frequent, rapid, and multiple "coalescence" of secondary neurospheres. (**a–f**) Representative sequential frames from a time-lapse video microscopic recording, show 30 spheres at the beginning of the recording which "merge" with each other (21 mergers counted), resulting in 10 spheres within ~10 h, and for some clusters (for example, in the *upper panels*) within 1 h. "Merger" partners are *circled* using *different colors* to facilitate following their movements and changes over the 10 h of the movie. These cultures were not agitated or otherwise manipulated; the movement reflects the intrinsic locomotion of free-floating spheres. Scale bar, 100 μm (with permission from Singec I et al. Nat Methods. 2006 Oct;3(10):801–6. Defining the actual sensitivity and specificity of the neurosphere assay in stem cell biology)

Another concern with the neuropshere assay is the potential of serial neurosphere passages, which require extended periods of cell culture, to induce significant alterations in cellular biology and gene expression that do not accurately reflect in vivo conditions [1]. Additionally, immunocytochemistry in neurospheres is challenging due to their size, fragility, and floating condition [48]. Cryostat sectioning for neurospheres gives the best reported results without disturbing the spherical architecture [49]. Prior to the sectioning, the neuropsheres need to be suspended in OCT compound and placed on the top of a frozen OCT mold to freeze [48].

Matrigel-Based Assays

Matrigel-based assays have been used as in vitro assays to measure glioma stem cell invasiveness [50]. Matrigel is a solubilized basement membrane preparation extracted from the Engelbreth–Holm–Swarm mouse sarcoma. This tumor is rich in ECM proteins and contains all the major components of basement membranes [51]. The Matrigel assay can be used to test the various molecular mechanisms that govern the invasion and migration of the TSC.

Qui et al., investigated the invasive potential of glioma stem cells and found that they were significantly more invasive than their differentiated progeny cells [52]. The same group also used this model to show that glioma stem cell invasiveness is markedly decreased after IL-6 is blocked with neutralizing antibody, but significantly increased when exogenous IL-6 is added [53].

Inoue and colleagues isolated tumor sphere-forming cells from U251 cells and showed enhanced migratory and invasive ability on both Matrigel and organotypic brain slices compared to parental cells. Furthermore, knockdown of MMP-13 expression by shRNA was shown to suppress the migration and invasion of the glioma stem cells [54]. Matrigel invasion assay have also been used to study neural and mesenchymal stem cell tropism to malignant glioma [55, 56].

The utility of Matrigel assays is limited however, as they do not reflect the human brain matrix. Other in vitro invasion assays such as wound healing assay [57], microliter-scale migration assay [58], spot assay [59], and transwell migration assays [60] have the same limitations.

Organotypic Cultures

In the context of an invasion assay, the neurosphere culture does not account for the novel ECM configuration of the brain along with cell-to-matrix interactions within a tumor [60–68]. To address this issue, organotypic slice cultures have been developed (Fig. 3). The early slice cultures were derived from postnatal rodent brain and have been widely used due to their easy access for pharmacological intervention and live imaging [69, 70]. Ohnishi used rat brain slices obtained from the hippocampus or cortical regions of 2-day-old rats and maintained the brain slices in culture at the interface between air and the culture medium [55].

Juvenile rat and mouse brain slice cultures can be kept viable for more than 8 weeks. Brain slice cultures have been developed using human tissue as well. Jung et al. used normal brain tissue specimens obtained from patients undergoing temporal lobectomies as a matrix to study glioma cell invasion [71]. Similarly, Chaichana and colleagues maintained intraoperatively collected human tumor and non-tumor explants ex vivo for approximately 11 days without any significant changes to the tissue cytoarchitecture [72].

Merz and colleagues described the use of ex vivo GBM slices in 6-well plates, and applied chemotherapeutic agents and irradiation, all the while allowing direct observation of the tumor response [70]. Similarly, Shimizu and colleagues have described a three-dimensional organotypic ex vivo system of surgical GBM specimens that preserves tumor cells in their original milieu [73]. Using this model, the group also described how Notch inhibition in explants resulted in decreased proliferation and self-renewal of tumor cells [74].

Organotypic coculture models are also valuable to study invasiveness of tumor cells [63, 64, 75, 76]. Aaberg-Jessen and colleagues implanted neurospheres derived from primary brain tumors into organotypic rat brain slice cultures and followed the

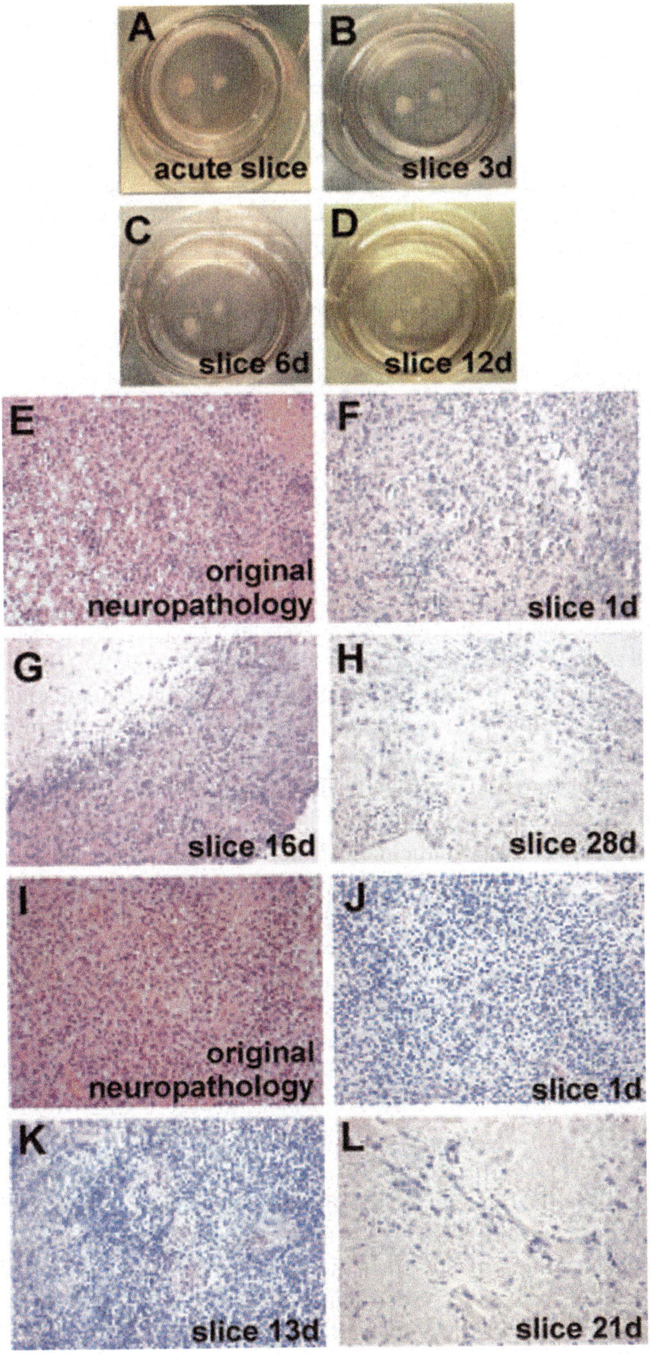

Fig. 3 Human GBM slices in culture. Slices were cultured on membrane inserts in six-well plates with no signs of degeneration\in acute (**a**) slices at 1 day or at 3 days (**b**), 6 days (**c**), or 12 days (**d**) in vitro. Original H and E neuropathology (**e** and **i**) and H and E-stained paraffin-embedded sections (8 mm; **f–h** and **j–l**) prepared from slices after various culture periods. Two different tumors (**e–h** and **i–l**) are shown. Note that typical features of individual tumors were maintained at least from 1 to 16 days (**f–g**) and 1–13 days (**j–k**) in vitro; massive cell loss was observed after 20 days in vitro (**h** and **l**). Original magnification: 1× in **a–d**; 200× in **e–l** (with permission from Merz et al. Neuro Oncol. 2013 Jun;15(6):670–81. Organotypic slice cultures of human glioblastoma reveal different susceptibilities to treatments)

invasive behavior of the tumor cells over time using confocal microscopy [77]. Likewise, Zhu et al. cocultured human brain microvascular endothelial cells (hBMEC) with GBM neurospheres and found increased NOTCH expression in endothelial cells leads to increased GBM cell growth and increased TSC self-renewal [78].

While the organotypic model has the advantages of more sophisticated cell-to-cell and cell-to-matrix interactions compared to other culture methods, there are a few limitations of the model that need to be considered. Although the vascular structures are preserved in the model, there is no blood flow along with nutrients, drugs or immune cells being carried through them. This limits its ability to test drugs that target blockage of the circulation system or modulation of the immune system [73].

Mathematical Models

Mathematical modeling represents a novel modality to develop predictive models for the biological behavior and treatment-response of TSC. Steel and colleagues accomplished some of the earliest work in mathematical modeling for gliomas [79]. Ganguly and Puri later formulated a predictive mechanistic mathematical model for brain TSC using a compartmental model [80]. Their group also formulated a model to understand the response to treatment of tumors with cancer stem cells. Their model predicted that the best response to chemotherapy occurs when a drug targets the abnormal stem cells [81].

A simple compartmental mathematical model for tumor growth, based on the TSC hypothesis using a chemical reaction approach has also been described [82]. Others have used multi-compartment models to predict growth potential of tumors with a heterogeneous cell population [83]. Mathematical models have also suggested that treatment modalities stimulating TSC differentiation and inhibiting TSC proliferation should be used together to get the best response [84].

Animal Models

Self-renewal and lineage capacity are the distinguishing features of any stem cell and any assay that aims to identify TSC needs the potential to show these characteristics. Serial transplantation in animal models is considered the best functional assay for these critical features [85–88]. In transplantation assays, tumor cells are xenografted into immunocompromised mice. Studies have typically used NOD/SCID mice [2], but BALB/c-nude [89] and Scid/bg mice [9] have also been used. The implanted tumors can then be assayed at various time points for tumor formation analysis.

To show self-renewal, the tumor cells need to be removed from the primary mice and transplanted into a secondary recipient animal. In the CNS, Singh and colleagues showed that only CD133+ brain tumor fraction contained cells that were capable of tumor initiation in vivo. They used an in vivo limiting dilution assay, where progressively smaller numbers of tumor cells are implanted into the animals to demonstrate the minimum number of cells required to form a tumor. Injection of as few as 100 CD133+ cells produced a tumor in vivo, while orthotopic implantation of 10^5 CD133− cells did not [90]. Likewise, Galli et al. demonstrated that glioblastoma cell lines, established by culture in neurosphere conditions, showed TSC characteristics in vivo. Intracranial injection of 200,000 of neurosphere cells also generated tumors in vivo, and after repeat culture, initiated phenotypically similar tumors, in a secondary mouse [9].

Xenotransplantation may also be employed to investigate the tumorigenicity of serially passaged neurospheres. Utilizing an athymic nude mouse model our group implanted 50,000 neurospheres into the basal ganglia, and after 4 weeks observed the generation of diffuse tumor in the basal ganglia depicting tumorigenic nature of the implanted cells [38]. To be able to follow the growth of the implanted TSC, Lathia and colleagues described the used of serial in vivo intravital microscopy [91]. To compare the role of TSC and non-stem tumor cells in the growth of the xenotransplanted tumors, GBM specimen cells were transduced with a lentivirus to express fluorescent proteins; green or yellow fluorescent protein for TSC, and cyan fluorescent protein for non-stem tumor cells. Additionally high-molecular weight fluorescent dextran injected to highlight the vessels around the tumor. Using real time imaging, the study was able to show that TSC were primarily responsible for the propagation of the implanted tumors [91] (Fig. 4).

Interestingly, glioma stem cells are not unique to xenotransplanted human cancers. Wu and colleagues were able to isolate CD133+ cells from the GL261 cell line, a syngeneic mouse glioma model [92]. CD133+ GL261 cells expressed nestin, formed tumor spheres with high frequency, and differentiated into glial and neuronal-like cells. Furthermore, a much lower number of murine CD133+ cells were needed to initiate tumors on intracerebral implantation compared to CD133− cells, 100 vs. 10,000 cells, respectively [92].

Although nude mice serial transplantation assays are considered the gold standard to identify and enrich stem cells, there are still some issues regarding the interpretation of the experiments. The effect of removing TSC from the supporting matrix is not known, and might modify characteristics of the cells [2]. Additionally the presence of predisposing genetic mutations that give rise to mouse tumors may fail to adequately represent the heterogeneity in human cancers [93]

It may also not possible to estimate the proportion of TSC in the parent tumor based on the results from animal transplantation assays [93]. The site and type of host tissue also has an important effect on the TSC representation. Vascularization at the site of implantation, extracellular matrix constitution, growth factor availability and host immunocompetence are some of the factors that can effect tumor engraftment and the yield of TSC [94].

Recently, Zebrafish (*Danio rerio*) assays have also been used to study the behavior of CNSTSC in vivo [95, 96]. As a vertebrate animal, the zebrafish model depicts high levels of physiologic and genetic similarities to mammals [96]. The transparent embryos of zebrafish are inexpensive to maintain and allow easy visualization of

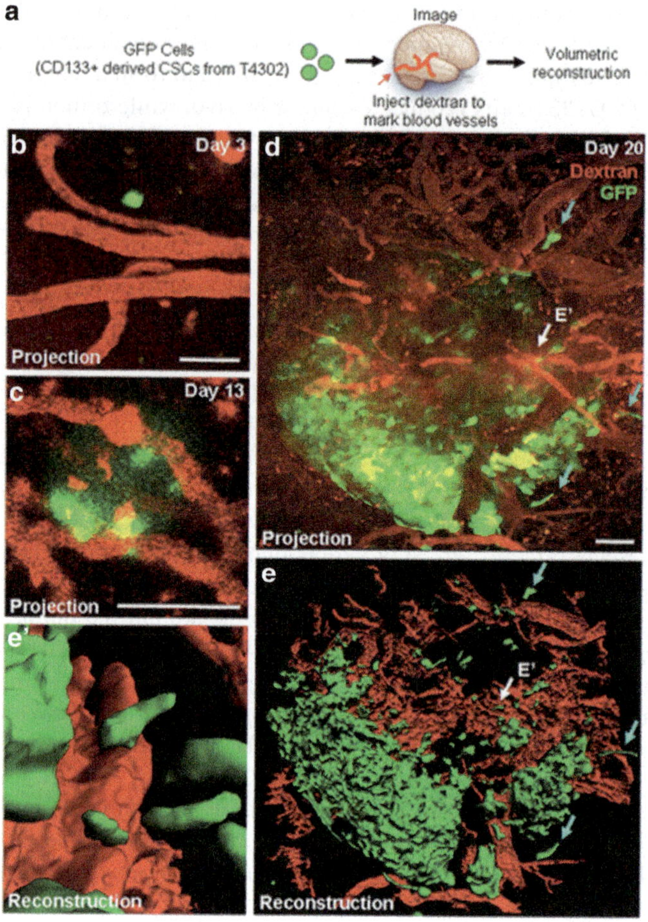

Fig. 4 Multiphoton microscopy reveals tumor propagation from cancer stem cells. Tumor formation in a xenotransplantation model was observed from GFP-labeled TSCs over time as shown in experimental design schematic (**a**). Projection micrographs (**b–d**) demonstrate tumor formation over time and three-dimensional reconstructions depicted in micrographs (**e**, **e**′) revealed tumor cells were closely associated with blood vessels (**e** shown with *white arrows* in **d**, **e**) and in peripheral areas (**d**, **e**, shown in *blue arrows*). Fluorescent dextran (shown in *red*) was injected into the circulation to illuminate blood vessels prior to imaging. Scale bar represents 50 μm (from Lathia JD et al. PLoS One. 2011;6(9):e24807. Direct in vivo evidence for tumor propagation by glioblastoma cancer stem cells (open access))

internal structures. A transparent adult zebrafish model has also been created to allow direct visualization of tumor engraftment and proliferation [97]. Investigators have used various imaging techniques to track the growth, invasiveness and response to experimental therapy, including fluorescence, bioluminescence and luciferase-based assays [96, 98, 99].

There are also some limitations of the zebrafish implantation model. The high mortality rate (>10 %) even with sublethal doses of radiation, poses a challenge for investigators. Post-transplant care also has to be optimized to minimize the risk of infection in the immunosuppressed recipients. Additional research is also required to examine the effect of background genotype on the behavior of the transplanted tumors, and to develop the ideal transgenic strain for particular tumors. Furthermore, injection of tumor cells in this model is mostly in the peritoneal cavity and while this may be a good model for metastatic tumors, the development of tissue-specific orthotropic injections will allow a more representative assay [100].

Conclusion

Increasing evidence points to a fundamental role for TSC in the initiation and propagation of several tumors. In the context of CNS tumors the development of treatment strategies specifically targeted towards TSC may hold a significant therapeutic promise. We have described some of the most commonly used laboratory models to investigate TSC to further this aim.

References

1. Rich JN, Eyler CE. Cancer stem cells in brain tumor biology. Cold Spring Harb Symp Quant Biol. 2008;73:411–20.
2. Clarke MF, Dick JE, Dirks PB, Eaves CJ, Jamieson CHM, Jones DL, et al. Cancer stem cells–perspectives on current status and future directions: AACR workshop on cancer stem cells. Cancer Res. 2006;66(19):9339–44.
3. Lapidot T, Sirard C, Vormoor J, Murdoch B, Hoang T, Caceres-Cortes J, et al. A cell initiating human acute myeloid leukaemia after transplantation into SCID mice. Nature. 1994;367 (6464):645–8.
4. Bonnet D, Dick JE. Human acute myeloid leukemia is organized as a hierarchy that originates from a primitive hematopoietic cell. Nat Med. 1997;3(7):730–7.
5. Taylor MD, Poppleton H, Fuller C, Su X, Liu Y, Jensen P, et al. Radial glia cells are candidate stem cells of ependymoma. Cancer Cell. 2005;8(4):323–35.
6. Ignatova TN, Kukekov VG, Laywell ED, Suslov ON, Vrionis FD, Steindler DA. Human cortical glial tumors contain neural stem-like cells expressing astroglial and neuronal markers in vitro. Glia. 2002;39(3):193–206.
7. Hemmati HD, Nakano I, Lazareff JA, Masterman-Smith M, Geschwind DH, Bronner-Fraser M, et al. Cancerous stem cells can arise from pediatric brain tumors. Proc Natl Acad Sci U S A. 2003;100(25):15178–83.
8. Singh SK, Clarke ID, Terasaki M, Bonn VE, Hawkins C, Squire J, et al. Identification of a cancer stem cell in human brain tumors. Cancer Res. 2003;63(18):5821–8.
9. Galli R, Binda E, Orfanelli U, Cipelletti B, Gritti A, De Vitis S, et al. Isolation and characterization of tumorigenic, stem-like neural precursors from human glioblastoma. Cancer Res. 2004;64(19):7011–21.
10. Yuan X, Curtin J, Xiong Y, Liu G, Waschsmann-Hogiu S, Farkas DL, et al. Isolation of cancer stem cells from adult glioblastoma multiforme. Oncogene. 2004;23(58):9392–400.
11. Fan X, Eberhart CG. Medulloblastoma stem cells. J Clin Oncol. 2008;26(17):2821–7.

12. deCarvalho AC, Nelson K, Lemke N, Lehman NL, Arbab AS, Kalkanis S, et al. Gliosarcoma stem cells undergo glial and mesenchymal differentiation in vivo. Stem Cells. 2010;28(2): 181–90.
13. Rycaj K, Tang DG. Cancer stem cells and radioresistance. Int J Radiat Biol. 2014;7.
14. Bao S, Wu Q, McLendon RE, Hao Y, Shi Q, Hjelmeland AB, et al. Glioma stem cells promote radioresistance by preferential activation of the DNA damage response. Nature. 2006;444(7120):756–60.
15. Vinogradov S, Wei X. Cancer stem cells and drug resistance: the potential of nanomedicine. Nanomedicine. 2012;7(4):597–615.
16. Eyler CE, Rich JN. Survival of the fittest: cancer stem cells in therapeutic resistance and angiogenesis. J Clin Oncol. 2008;26(17):2839–45.
17. Beier D, Hau P, Proescholdt M, Lohmeier A, Wischhusen J, Oefner PJ, et al. CD133(+) and CD133(−) glioblastoma-derived cancer stem cells show differential growth characteristics and molecular profiles. Cancer Res. 2007;67(9):4010–5.
18. Wang J, Sakariassen PØ, Tsinkalovsky O, Immervoll H, Bøe SO, Svendsen A, et al. CD133 negative glioma cells form tumors in nude rats and give rise to CD133 positive cells. Int J Cancer. 2008;122(4):761–8.
19. Ogden AT, Waziri AE, Lochhead RA, Fusco D, Lopez K, Ellis JA, et al. Identification of A2B5+CD133− tumor-initiating cells in adult human gliomas. Neurosurgery. 2008;62(2): 505–14. discussion 514–515.
20. Reynolds BA, Rietze RL. Neural stem cells and neurospheres–re-evaluating the relationship. Nat Methods. 2005;2(5):333–6.
21. Kuhn HG, Winkler J, Kempermann G, Thal LJ, Gage FH. Epidermal growth factor and fibroblast growth factor-2 have different effects on neural progenitors in the adult rat brain. J Neurosci. 1997;17(15):5820–9.
22. Craig CG, Tropepe V, Morshead CM, Reynolds BA, Weiss S, van der Kooy D. In vivo growth factor expansion of endogenous subependymal neural precursor cell populations in the adult mouse brain. J Neurosci. 1996;16(8):2649–58.
23. Rai KS, Hattiangady B, Shetty AK. Enhanced production and dendritic growth of new dentate granule cells in the middle-aged hippocampus following intracerebroventricular FGF-2 infusions. Eur J Neurosci. 2007;26(7):1765–79.
24. Yoshimura S, Takagi Y, Harada J, Teramoto T, Thomas SS, Waeber C, et al. FGF-2 regulation of neurogenesis in adult hippocampus after brain injury. Proc Natl Acad Sci U S A. 2001;98(10):5874–9.
25. Yoshimura S, Teramoto T, Whalen MJ, Irizarry MC, Takagi Y, Qiu J, et al. FGF-2 regulates neurogenesis and degeneration in the dentate gyrus after traumatic brain injury in mice. J Clin Invest. 2003;112(8):1202–10.
26. Nakano I, Kornblum HI. Methods for analysis of brain tumor stem cell and neural stem cell self-renewal. Methods Mol Biol. 2009;568:37–56.
27. Gage FH. Mammalian neural stem cells. Science. 2000;287(5457):1433–8.
28. Lie DC, Song H, Colamarino SA, Ming G, Gage FH. Neurogenesis in the adult brain: new strategies for central nervous system diseases. Annu Rev Pharmacol Toxicol. 2004;44: 399–421.
29. Vescovi AL, Galli R, Reynolds BA. Brain tumour stem cells. Nat Rev Cancer. 2006;6(6): 425–36.
30. Hermann A, Maisel M, Liebau S, Gerlach M, Kleger A, Schwarz J, et al. Mesodermal cell types induce neurogenesis from adult human hippocampal progenitor cells. J Neurochem. 2006;98(2):629–40.
31. Chaichana K, Zamora-Berridi G, Camara-Quintana J, Quiñones-Hinojosa A. Neurosphere assays: growth factors and hormone differences in tumor and nontumor studies. Stem Cells. 2006;24(12):2851–7.
32. Gritti A, Galli R, Vescovi AL. Clonal analyses and cryopreservation of neural stem cell cultures. Methods Mol Biol. 2008;438:173–84.
33. Clement V, Sanchez P, de Tribolet N, Radovanovic I, Ruiz i Altaba A. HEDGEHOG-GLI1 signaling regulates human glioma growth, cancer stem cell self-renewal, and tumorigenicity. Curr Biol. 2007;17(2):165–72.

34. Ma Z, Lin M, Li K, Fu Y, Liu X, Yang D, et al. Knocking down SMC1A inhibits growth and leads to G2/M arrest in human glioma cells. Int J Clin Exp Pathol. 2013;6(5):862–9.
35. Li G, Wang R, Gao J, Deng K, Wei J, Wei Y. RNA interference-mediated silencing of iASPP induces cell proliferation inhibition and G0/G1 cell cycle arrest in U251 human glioblastoma cells. Mol Cell Biochem. 2011;350(1–2):193–200.
36. George J, Banik NL, Ray SK. Knockdown of hTERT and concurrent treatment with interferon-gamma inhibited proliferation and invasion of human glioblastoma cell lines. Int J Biochem Cell Biol. 2010;42(7):1164–73.
37. Guo G, Kuai D, Cai S, Xue N, Liu Y, Hao J, et al. Knockdown of FRAT1 expression by RNA interference inhibits human glioblastoma cell growth, migration and invasion. PLoS One. 2013;8(4):e61206.
38. Ehtesham M, Mapara KY, Stevenson CB, Thompson RC. CXCR4 mediates the proliferation of glioblastoma progenitor cells. Cancer Lett. 2009;274(2):305–12.
39. Ehtesham M, Kabos P, Gutierrez MAR, Chung NHC, Griffith TS, Black KL, et al. Induction of glioblastoma apoptosis using neural stem cell-mediated delivery of tumor necrosis factor-related apoptosis-inducing ligand. Cancer Res. 2002;62(24):7170–4.
40. Fael Al-Mayhani TM, Ball SLR, Zhao J-W, Fawcett J, Ichimura K, Collins PV, et al. An efficient method for derivation and propagation of glioblastoma cell lines that conserves the molecular profile of their original tumours. J Neurosci Methods. 2009;176(2):192–9.
41. Snyder EY, Deitcher DL, Walsh C, Arnold-Aldea S, Hartwieg EA, Cepko CL. Multipotent neural cell lines can engraft and participate in development of mouse cerebellum. Cell. 1992;68(1):33–51.
42. Parker MA, Anderson JK, Corliss DA, Abraria VE, Sidman RL, Park KI, et al. Expression profile of an operationally-defined neural stem cell clone. Exp Neurol. 2005;194(2):320–32.
43. Suslov ON, Kukekov VG, Ignatova TN, Steindler DA. Neural stem cell heterogeneity demonstrated by molecular phenotyping of clonal neurospheres. Proc Natl Acad Sci U S A. 2002;99(22):14506–11.
44. Kukekov VG, Laywell ED, Thomas LB, Steindler DA. A nestin-negative precursor cell from the adult mouse brain gives rise to neurons and glia. Glia. 1997;21(4):399–407.
45. Singec I, Knoth R, Meyer RP, Maciaczyk J, Volk B, Nikkhah G, et al. Defining the actual sensitivity and specificity of the neurosphere assay in stem cell biology. Nat Methods. 2006;3(10):801–6.
46. Reynolds BA, Tetzlaff W, Weiss S. A multipotent EGF-responsive striatal embryonic progenitor cell produces neurons and astrocytes. J Neurosci. 1992;12(11):4565–74.
47. Reynolds BA, Weiss S. Clonal and population analyses demonstrate that an EGF-responsive mammalian embryonic CNS precursor is a stem cell. Dev Biol. 1996;175(1):1–13.
48. Guerrero-Cázares H, Chaichana KL, Quiñones-Hinojosa A. Neurosphere culture and human organotypic model to evaluate brain tumor stem cells. Methods Mol Biol. 2009;568: 73–83.
49. Parmar M, Sjöberg A, Björklund A, Kokaia Z. Phenotypic and molecular identity of cells in the adult subventricular zone in vivo and after expansion in vitro. Mol Cell Neurosci. 2003;24(3):741–52.
50. Yu S, Yang X, Zhang B, Ming H, Chen C, Ren B, et al. Enhanced invasion in vitro and the distribution patterns in vivo of CD133+ glioma stem cells. Chin Med J (Engl). 2011; 124(17):2599–604.
51. M D Brooks SA. In vitro invasion assay using Matrigel®. In: Brooks SA, Schumacher U, editors. Metastasis research protocols. Humana Press; p. 61–70. http://link.springer.com/protocol/10.1385/1-59259-137-X%3A061. Accessed 20 Jun 2013.
52. Qiu B, Zhang D, Tao J, Tie X, Wu A, Wang Y. Human brain glioma stem cells are more invasive than their differentiated progeny cells in vitro. J Clin Neurosci. 2012;19(1):130–4.
53. Qiu B, Zhang D, Wang Y, Ou S, Wang J, Tao J, et al. Interleukin-6 is overexpressed and augments invasiveness of human glioma stem cells in vitro. Clin Exp Metastasis. 2013; 30(8):1009–18.
54. Inoue A, Takahashi H, Harada H, Kohno S, Ohue S, Kobayashi K, et al. Cancer stem-like cells of glioblastoma characteristically express MMP-13 and display highly invasive activity. Int J Oncol. 2010;37(5):1121–31.

55. Koizumi S, Gu C, Amano S, Yamamoto S, Ihara H, Tokuyama T, et al. Migration of mouse-induced pluripotent stem cells to glioma-conditioned medium is mediated by tumor-associated specific growth factors. Oncol Lett. 2011;2(2):283–8.
56. Hata N, Shinojima N, Gumin J, Yong R, Marini F, Andreeff M, et al. Platelet-derived growth factor BB mediates the tropism of human mesenchymal stem cells for malignant gliomas. Neurosurgery. 2010;66(1):144–56. discussion 156–157.
57. Nobes CD, Hall A. Rho GTPases control polarity, protrusion, and adhesion during cell movement. J Cell Biol. 1999;144(6):1235–44.
58. McDonough W, Tran N, Giese A, Norman SA, Berens ME. Altered gene expression in human astrocytoma cells selected for migration: I. Thromboxane synthase. J Neuropathol Exp Neurol. 1998;57(5):449–55.
59. Wichterle H, Alvarez-Dolado M, Erskine L, Alvarez-Buylla A. Permissive corridor and diffusible gradients direct medial ganglionic eminence cell migration to the neocortex. Proc Natl Acad Sci U S A. 2003;100(2):727–32.
60. Albini A, Iwamoto Y, Kleinman HK, Martin GR, Aaronson SA, Kozlowski JM, et al. A rapid in vitro assay for quantitating the invasive potential of tumor cells. Cancer Res. 1987;47(12):3239–45.
61. Amar AP, DeArmond SJ, Spencer DR, Coopersmith PF, Ramos DM, Rosenblum ML. Development of an in vitro extracellular matrix assay for studies of brain tumor cell invasion. J Neurooncol. 1994;20(1):1–15.
62. Bernstein JJ, Goldberg WJ, Laws Jr ER, Conger D, Morreale V, Wood LR. C6 glioma cell invasion and migration of rat brain after neural homografting: ultrastructure. Neurosurgery. 1990;26(4):622–8.
63. Bjerkvig R, Laerum OD, Mella O. Glioma cell interactions with fetal rat brain aggregates in vitro and with brain tissue in vivo. Cancer Res. 1986;46(8):4071–9.
64. Bjerkvig R, Tønnesen A, Laerum OD, Backlund EO. Multicellular tumor spheroids from human gliomas maintained in organ culture. J Neurosurg. 1990;72(3):463–75.
65. Chicoine MR, Silbergeld DL. Invading C6 glioma cells maintaining tumorigenicity. J Neurosurg. 1995;83(4):665–71.
66. Erkell LJ, Schirrmacher V. Quantitative in vitro assay for tumor cell invasion through extracellular matrix or into protein gels. Cancer Res. 1988;48(23):6933–7.
67. Kramer RH, Bensch KG, Wong J. Invasion of reconstituted basement membrane matrix by metastatic human tumor cells. Cancer Res. 1986;46(4 Pt 2):1980–9.
68. Pilkington GJ, Bjerkvig R, De Ridder L, Kaaijk P. In vitro and in vivo models for the study of brain tumour invasion. Anticancer Res. 1997;17(6B):4107–9.
69. Gähwiler BH. Slice cultures of cerebellar, hippocampal and hypothalamic tissue. Experientia. 1984;40(3):235–43.
70. Merz F, Gaunitz F, Dehghani F, Renner C, Meixensberger J, Gutenberg A, et al. Organotypic slice cultures of human glioblastoma reveal different susceptibilities to treatments. Neuro Oncol. 2013;15(6):670–81.
71. Jung S, Kim H-W, Lee J-H, Kang S-S, Rhu H-H, Jeong Y-I, et al. Brain tumor invasion model system using organotypic brain-slice culture as an alternative to in vivo model. J Cancer Res Clin Oncol. 2002;128(9):469–76.
72. Chaichana KL, Capilla-Gonzalez V, Gonzalez-Perez O, Pradilla G, Han J, Olivi A, et al. Preservation of glial cytoarchitecture from ex vivo human tumor and non-tumor cerebral cortical explants: a human model to study neurological diseases. J Neurosci Methods. 2007;164(2):261–70.
73. Shimizu F, Hovinga KE, Metzner M, Soulet D, Tabar V. Organotypic explant culture of glioblastoma multiforme and subsequent single-cell suspension. Curr Protoc Stem Cell Biol. 2011;Chapter 3:Unit3.5.
74. Hovinga KE, Shimizu F, Wang R, Panagiotakos G, Van Der Heijden M, Moayedpardazi H, et al. Inhibition of notch signaling in glioblastoma targets cancer stem cells via an endothelial cell intermediate. Stem Cells. 2010;28(6):1019–29.
75. Pedersen PH, Edvardsen K, Garcia-Cabrera I, Mahesparan R, Thorsen J, Mathisen B, et al. Migratory patterns of lac-z transfected human glioma cells in the rat brain. Int J Cancer. 1995;62(6):767–71.

76. Engebraaten O, Bjerkvig R, Lund-Johansen M, Wester K, Pedersen PH, Mørk S, et al. Interaction between human brain tumour biopsies and fetal rat brain tissue in vitro. Acta Neuropathol. 1990;81(2):130–40.
77. Aaberg-Jessen C, Nørregaard A, Christensen K, Pedersen CB, Andersen C, Kristensen BW. Invasion of primary glioma- and cell line-derived spheroids implanted into corticostriatal slice cultures. Int J Clin Exp Pathol. 2013;6(4):546–60.
78. Zhu TS, Costello MA, Talsma CE, Flack CG, Crowley JG, Hamm LL, et al. Endothelial cells create a stem cell niche in glioblastoma by providing NOTCH ligands that nurture self-renewal of cancer stem-like cells. Cancer Res. 2011;71(18):6061–72.
79. Steel GG. Growth kinetics of tumours: cell population kinetics in relation to the growth and treatment of cancer. England: Clarendon; 1977. p. 380.
80. Ganguly R, Puri IK. Mathematical model for the cancer stem cell hypothesis. Cell Prolif. 2006;39(1):3–14.
81. Ganguly R, Puri IK. Mathematical model for chemotherapeutic drug efficacy in arresting tumour growth based on the cancer stem cell hypothesis. Cell Prolif. 2007;40(3):338–54.
82. Molina-Peña R, Álvarez MM. A simple mathematical model based on the cancer stem cell hypothesis suggests kinetic commonalities in solid tumor growth. PLoS One. 2012;7(2):e26233.
83. Weekes SL, Barker B, Bober S, Cisneros K, Cline J, Thompson A, et al. A multicompartment mathematical model of cancer stem cell-driven tumor growth dynamics. Bull Math Biol. 2014;76(7):1762–82.
84. Vainstein V, Kirnasovsky OU, Kogan Y, Agur Z. Strategies for cancer stem cell elimination: insights from mathematical modeling. J Theor Biol. 2012;298:32–41.
85. Reya T, Morrison SJ, Clarke MF, Weissman IL. Stem cells, cancer, and cancer stem cells. Nature. 2001;414(6859):105–11.
86. Pardal R, Clarke MF, Morrison SJ. Applying the principles of stem-cell biology to cancer. Nat Rev Cancer. 2003;3(12):895–902.
87. Dick JE. Breast cancer stem cells revealed. Proc Natl Acad Sci U S A. 2003;100(7):3547–9.
88. Tysnes BB. Tumor-initiating and -propagating cells: cells that we would like to identify and control. Neoplasia N Y N. 2010 Jul;12(7):506–15.
89. Choi SA, Lee JY, Phi JH, Wang K-C, Park C-K, Park S-H, et al. Identification of brain tumour initiating cells using the stem cell marker aldehyde dehydrogenase. Eur J Cancer. 2014;50(1):137–49.
90. Singh SK, Hawkins C, Clarke ID, Squire JA, Bayani J, Hide T, et al. Identification of human brain tumour initiating cells. Nature. 2004;432(7015):396–401.
91. Lathia JD, Gallagher J, Myers JT, Li M, Vasanji A, McLendon RE, et al. Direct in vivo evidence for tumor propagation by glioblastoma cancer stem cells. PLoS One. 2011;6(9):e24807.
92. Wu A, Oh S, Wiesner SM, Ericson K, Chen L, Hall WA, et al. Persistence of CD133+ cells in human and mouse glioma cell lines: detailed characterization of GL261 glioma cells with cancer stem cell-like properties. Stem Cells Dev. 2008;17(1):173–84.
93. Kelly PN, Dakic A, Adams JM, Nutt SL, Strasser A. Tumor growth need not be driven by rare cancer stem cells. Science. 2007;317(5836):337.
94. Gupta PB, Chaffer CL, Weinberg RA. Cancer stem cells: mirage or reality? Nat Med. 2009;15(9):1010–2.
95. Dovey MC, Zon LI. Defining cancer stem cells by xenotransplantation in zebrafish. Methods Mol Biol. 2009;568:1–5.
96. Yang X-J, Cui W, Gu A, Xu C, Yu S-C, Li T-T, et al. A novel zebrafish xenotransplantation model for study of glioma stem cell invasion. PLoS One. 2013;8(4):e61801.
97. White RM, Sessa A, Burke C, Bowman T, LeBlanc J, Ceol C, et al. Transparent adult zebrafish as a tool for in vivo transplantation analysis. Cell Stem Cell. 2008;2(2):183–9.
98. Zhao H, Tang C, Cui K, Ang B-T, Wong STC. A screening platform for glioma growth and invasion using bioluminescence imaging. J Neurosurg. 2009;111(2):238–46.
99. Sato A, Klaunberg B, Tolwani R. In vivo bioluminescence imaging. Comp Med. 2004;54(6):631–4.
100. Taylor AM, Zon LI. Zebrafish tumor assays: the state of transplantation. Zebrafish. 2009;6(4):339–46.

Radiation Therapy for Glioma Stem Cells

Anthony E. Rizzo and Jennifer S. Yu

Abstract Radiation therapy is the most effective adjuvant treatment modality for virtually all patients with high-grade glioma. Its ability to improve patient survival has been recognized for decades. Cancer stem cells provide new insights into how tumor biology is affected by radiation and the role that this cell population can play in disease recurrence. Glioma stem cells possess a variety of intracellular mechanisms to resist and even flourish in spite of radiation, and their proliferation and maintenance appear tied to supportive stimuli from the tumor microenvironment. This chapter reviews the basis for our current use of radiation to treat high-grade gliomas, and addresses this model in the context of therapeutically resistant stem cells. We discuss the available evidence highlighting current clinical efforts to improve radiosensitivity, and newer targets worthy of further development.

Keywords Glioma • Stem cell • Initiating cell • Glioblastoma • Radiation • Radioresistance • Microenvironment • Hypoxia • DNA damage repair

Radiation therapy is a cornerstone in the treatment of a variety of primary brain tumors. The identification and study of cancer stem cells in a number of primary CNS neoplasms provides new insights into how tumor biology is affected by radiation and the role that cancer stem cells can play in therapeutic resistance and recurrence of some devastating diseases. While the WHO classification system lists numerous CNS tumors, studies identifying and characterizing cancer stem cells are present only in a limited number of them. This chapter focuses on the evidence

---

A.E. Rizzo, B.A.
Department of Stem Cell Biology and Regenerative Medicine, Lerner Research Institute, Cleveland Clinic, 9500 Euclid Avenue, Cleveland, OH 44195, USA
e-mail: rizzoa4@ccf.org

J.S. Yu, M.D., Ph.D. (✉)
Department of Radiation Oncology, Cleveland Clinic, Cleveland, OH 44124, USA

Department of Stem Cell Biology and Regenerative Medicine, Lerner Research Institute, Cleveland Clinic, 9500 Euclid Avenue, Cleveland, OH 44195, USA
e-mail: yuj2@ccf.org

M. Ehtesham (ed.), *Stem Cell Biology in Neoplasms of the Central Nervous System*, Advances in Experimental Medicine and Biology 853,
DOI 10.1007/978-3-319-16537-0_6

gathered in the study of cancer stem cells related to gliomas. These tumors represent the most common malignant cancers of the CNS and thus a disproportionate amount of investigation is focused on this disease burden. Lessons from glioma may serve as a model for other CNS tumors for which stem cell populations have been identified.

Comprising four different tumor grades, astrocytoma are the most common primary brain tumors [1]. The most common high-grade astrocytoma is glioblastoma (GBM, WHO grade IV astrocytoma) which accounts for 45 % of all primary malignant CNS tumors and carries the gravest prognosis [2]. Despite decades of significant research efforts, the median survival for patients with newly diagnosed GBM receiving the standard of care is between 12 and 15 months, with a 5-year survival rate of approximately 3 % [3]. Optimal therapy for this highly invasive and therapeutically resistant disease includes maximal safe surgical resection [4], radiation therapy [5, 6], and chemotherapy [3]. While surgical resection has been found to provide a survival benefit, less than half of the patients with newly diagnosed disease are candidates for the optimal procedure, gross total resection, and virtually all GBMs will recur following surgery [4, 7]. Radiation therapy is the most effective adjuvant treatment modality for virtually all GBM patients and its ability to improve patient overall survival has been recognized for some time [8, 9].

GBM has long been regarded as a relatively radioresistant cancer, in part due to the persistently high failure rates (up to 90 %) in patients treated with radiation doses up to 80 Gy [10]. Though recurrence may only be delayed with current treatments, the efficacy of adjuvant radiation therapy to improve survival has been supported by level I clinical evidence from randomized controlled trials since the 1970s [8, 9, 11]. The current standard practice of radiation therapy used to treat GBM is based on clinical studies done over the past 40 years. Prospective randomized trials established that whole-brain irradiation did not yield improvements in overall survival or changes in recurrence pattern compared to partial brain irradiation [12, 13]. The practice of delivering 60 Gy to the gross tumor volume is based on a pooled analysis of three successive randomized trials in which a progressive increase in median overall survival was observed among doses ranging from less than 45 to 60 Gy [5]. A dose-effect relationship was suggested for doses above 50 Gy by this study, and following studies comparing a dose of 70–60 Gy demonstrated no survival or local control advantage, which established the dose of 60 Gy [14, 15].

The practice of treating the gross tumor volume and a margin of 2 cm added to the FLAIR and T1-enhancing regions is based on studies of recurrence patterns and microinvasive disease. Early studies established that almost 90 % of recurrences occur within 2 cm of the primary tumor site when assessed by computed tomography (CT) [16, 17]. Magnetic resonance imaging (MRI), specifically the gadolinium-enhanced T1 imaging sequence, is more sensitive than CT in defining tumor extension [18–20]. However, biopsy and autopsy studies further demonstrated that microinvasive disease can be present within the 1–4 cm margins of the gross tumor volume defined by T1-contrast enhancement [21]. More recent retrospective analyses, which evaluate patients that received the current standard of chemoradiotherapy, continue to show a large proportion of recurrence centrally and within the radiation treatment field [22].

The brain tumor stem cell theory as a model for GBM tumorigenesis, therapeutic resistance, and recurrence is still under relatively contentious debate [23]. Despite its continued development and testing, this model has important prognostic and therapeutic implications [24, 25]. For the purposes of this chapter, we use the term GBM-initiating cell (GIC) to refer to a population of cells that are alternately described in the literature as "GBM stem cell" or "brain tumor stem-like cell." These terms are united by a functional definition that is in accordance with the cancer stem cell hypothesis [26]. GICs are a relatively small subpopulation of cells, isolated from bulk tumor specimens, that are experimentally defined by the capacities for self-renewal, differentiation, and maintenance of proliferation [27]. The term GIC is not meant to suggest that this cell population is necessarily the origin of gliomas. Several groups have described GICs using validated functional assays such as serial tumorsphere assay and tumor propagation by in vivo intracranial limiting dilution assay [28]. While a number of cell surface markers have been used to aid in the characterization and isolation of GICs there is no clear universal marker for GICs [29]. This may be due in part to inherent differences between genotypes in human tumors or proposed plasticity of the GIC phenotype [30–33]. Most likely, several different populations of GICs exist within a tumor and each may express different combinations of cell surface markers. Most significant for the assessment of radiation therapy in the treatment of GBM is that GICs display much greater tumorigenic potential than matched non-GIC tumor cells when xenotransplanted into the brains of immunocompromised rodents [34]. Thus, treatments targeting GICs are attractive goals for reducing the recurrence of GBM.

Reflecting on the high propensity for local recurrence of GBM in the face of currently clinically optimized therapeutic practice yields some general questions for further discussion in the context of GICs. First, is radiation an effective treatment for GICs? It would appear clear from prior data that radiation is effective at reducing the tumor bulk, but what about its efficacy on GICs? Second, how do GICs respond to radiation? Clearly there is some population of tumor cells that are not eradicated by current practice; if they are GICs as we currently understand them, can any differences be exploited therapeutically? And finally, is radiation targeting these cells effectively? Can radiation shift the balance between GICs and non-GICs or cause non-GICs to adopt the GIC phenotype?

Ionizing radiation causes cell damage in a variety of ways; however the mechanism believed to be most responsible is the generation of reactive oxygen species leading to DNA damage in the form of double-strand breaks (DSB) [35]. The presence of DSBs, either induced exogenously or the result of endogenous forces during the cell cycle, represents potentially fatal obstacles for cells undergoing replication and division [36]. Unsurprisingly, the presence of DSB stimulates the activation of an array of proteins referred to as the DNA damage response (DDR), which is essential for cells to recover from DSBs [37]. The DDR encompasses a diverse but interconnected set of cellular processes including the damage sensors which initiate and transduce its signal and effectors that modulate cell cycle progression, DNA repair, autophagy, mitotic catastrophe, necrosis, senescence, and apoptosis [37]. The primary function of this cascade is the prevention of DSBs from being transmitted to later generations of cells as mutations or chromosomal aberrations [36]. This prevention can take the form of cell cycle arrest to allow for repair or organized cell destruction in the event of severe damage. It is important to note that it is thus not the actual insult of ionizing radiation that initiates destruction of cells, but rather the ensuing challenges to DNA replication and transcription. Accordingly, a variety of mechanisms ranging from the reduction of DSB generation, increased capacity for DNA repair, or overcoming cell cycle arrest or organized destruction are implicated in resistance to radiation therapy.

Consistent with its reputation as a radioresistant cancer, cell lines derived from gross specimens of GBM have been found to possess aberrant constitutive activation of a range of DDR proteins [38]. This would suggest that bulk tumor cells may be responsible for radioresistance and recurrence following radiation therapy. However, in response to ionizing radiation GICs are found to preferentially activate a number of critical components of the DDR (specifically ATM, Rad17, Chk2, and Chk1) and more efficiently repair DNA damage in comparison to matched non-GICs [39]. Irradiated GICs also have a lower percentage of apoptotic cells than matched non-GICs [39, 40]. Apoptosis, programmed cell death, is one protective mechanism activated by the DDR cascade in response to extensive DNA damage [35]. The ability to preferentially overcome cell cycle arrest and repair DNA damage compared to non-GIC bulk tumor cell places GICs in a position to expand their population and repopulate the tumor. Evidence in support of GICs as a source of therapeutic resistance can be found by studying tumor composition following radiation therapy. Expansion of the GIC population within recurrent tumors has been confirmed by histological analysis of GBM samples collected at the time of salvage surgery, after initial chemoradiation [41]. In addition, many clinical trials (though not all) have failed to show a benefit to radiation dose escalation [42], radiosurgery boost [43, 44], or brachytherapy boost [45, 46]. These data suggest that while radiation may be effective at reducing the tumor bulk by targeting non-GICs, it exerts a selective pressure favoring the outgrowth of an aggressive recurrent tumor through the expansion of the GIC population. Efficient repair of DSBs is one possible mechanism for the superior response to ionizing radiation seen in GICs, and while the exact cellular mechanisms responsible for this superior response are still being elucidated, we will endeavor to outline the current understanding.

It is important to distinguish that thus far, the study of radioresistance in GICs has not revealed an overexpression of proteins directly involved in the DDR cascade. GICs commonly have a basal activation or preferential activation of these proteins in response to ionizing radiation or possess mechanisms to overcome cell cycle arrest checkpoints [39, 47, 48]. Interestingly, the specific proteins that are preferentially activated and the degree to which they are activated vary between patient-derived GIC specimens, suggesting that there is not a single mechanism for radioresistance [49]. The cellular processes that support preferential DDR activation in response to radiation therapy could be indirectly stimulated by other aberrations that are specific or predominant in the maintenance and proliferation of a GIC population. One could further hypothesize that the resultant radioresistance of GICs is a corollary of signaling that supports the GIC phenotype.

A better understanding of the molecular response to radiation therapy can identify targets for therapies aimed at the specific cellular components. Realizing this potential involves dissecting the DDR cascade. This signaling cascade includes multiple sensor, transducer, and effector proteins. An important sensor of DNA damage is the MRE11-RAD50-NBS1 (MRN) complex [50, 51]. This complex rapidly binds to and assembles at foci of DNA damage and is an important activator of transducing proteins that continue the cascade [52]. Two important transducing proteins are the serine/threonine protein kinases ataxia telangiectasia mutated (ATM) and ataxia telangiectasia and Rad3-related protein (ATR) [53]. These two proteins are members of the phosphatidylinositol 3-kinase (PI3K) family and are key regulators of DSB repair. ATM and ATR each activates checkpoint effector proteins CHK2 and CHK1, respectively [53]. Each of these components may play a role in distinguishing the GIC response to ionizing radiation compared to non-GIC.

The involvement of the MRN complex as a potential starting point for preferential activation of DDR proteins is significant for GICs. An upstream regulator of the NBS1 protein in this complex is preferentially expressed in GICs. The cellular surface marker L1CAM is preferentially expressed on GICs to maintain cell survival and tumor growth [54]. Subsequent study of this marker identified a new function of L1CAM in promoting DDR checkpoint activation and radioresistance of GICs through regulation of NBS1 [55]. The L1CAM intracellular domain (L1-ICD) translocates to the nucleus in response to radiation, and it was found that knockdown of L1CAM reduced levels of the transcription factor c-Myc [55]. c-Myc is known to directly regulate NBS1 and is required for the ATM-dependent CHK2 activation, the downstream effector of the MRN complex in DDR [56, 57]. Taken together, the preferential expression of L1CAM provides a basis for the preferential activation of DDR proteins by GICs in response to ionizing radiation.

The transcription factor c-Myc is an important regulator of stem cell biology in both normal and cancer cells, and its transcriptional targets regulate proliferation, apoptosis, and malignant transformation [58]. The oncogenic potential of c-Myc was recognized in the 1980s, and further studies have demonstrated that overexpression of it correlates with poor prognosis in a variety of human tumors [59]. This transcription factor is found to not only be required for GIC proliferation, growth, and survival, but it is also expressed at higher levels in GICs compared to matched non-GICs [60]. Knockdown of c-Myc significantly reduced GIC growth and proliferation, and altered the expression of numerous cell cycle regulators downstream of c-Myc. This effect was seen preferentially in GICs compared to matched non-GICs, which were minimally disturbed [60]. The connection between L1CAM, c-Myc, and the DDR machinery illustrates one of the ways that radiation resistance appears tightly linked to the biology that defines GICs. Thus as a regulator of both stem cell maintenance and radioresistance L1CAM signaling represents an attractive target for modulation of radiation therapy.

DSBs are sensed by the MRN complex and ATM, and these sensors are interdependent for the recognition and signaling of DSBs. When both ATM and the MRN complex are recruited to the foci of DNA damage the MRN complex accelerates phosphorylation of inactive ATM dimers leading to their dissociation [51, 53, 61, 62].

Each phosphorylated ATM monomer further activates itself by auto-phosphorylation in a feed-forward mechanism to activate effector proteins including CHK2 kinase [63]. The CHK2 protein is a molecular switch which directly activates various targets. These include proteins involved in cell cycle progression, DNA repair, and stimulation of apoptosis. The second transducing protein, ATR, functions similarly to ATM, but predominantly in response to endogenous DNA damage. ATR may also be activated in response to DSBs induced by ionizing radiation, though to a lesser extent than ATM [64]. The signaling cascade downstream of ATR begins with activation of CHK1 [64]. CHK1 and CHK2 demonstrate over-lapping, but nonredundant, roles in their effects on cell cycle progression, DNA repair, and apoptosis [65]. One general distinction is the implication of ATM-CHK2 in the G1 checkpoint, with ATR-CHK1 having a more significant role in modulating the S- and G2-phase checkpoints [66].

While the direct contributions of the ATM-CHK2 and ATR-CHK1 remain unclear, several findings support continued investigation of inhibitors of these proteins as therapeutic targets. First, ATM-CHK2 is preferentially activated in GICs and inhibition of the CHK1/2 proteins improves GIC sensitivity to ionizing radiation [39]. Second, ATM expression correlates with radioresistance in bulk GBM cells [67], while its inhibitors have been found to increase the radiosensitivity of bulk GBM cells and GICs treated with temozolomide and radiation [68, 69]. Other targets for drug development can be found downstream of the CHK1/2 proteins. The ATM-CHK2 cascade activates transcription factors that alter the expression of numerous genes including the receptor tyrosine kinase c-MET, which has specific significance for GICs [70].

c-MET is a receptor tyrosine kinase with downstream targets involved in a variety of cellular signaling pathways including proliferation, motility, migration, and invasion [71]. c-MET is overexpressed in approximately 29 % of GBM patients and directly correlates with poor prognosis [72–77]. While the gene MET is amplified in only 5 % of GBM patients, its function is important in the context of GICs. Subpopulations of GBM cells enriched for elevated c-MET expression from primary GBM possess stem-like characteristics such as in vivo tumor initiation [78]. c-MET is activated after interaction with its ligand, hepatocyte growth factor/scatter factor (HGF/SF), which is secreted in an autocrine fashion by GICs [79]. This autocrine/paracrine loop is important for the maintenance of the GIC phenotype. Irradiation upregulates the expression of c-MET in GICs, highlighting the significance of this receptor and its potential to support recurrence following radiation therapy [78].

In response to ionizing radiation, c-MET expression and activation are increased, as is secretion of HGF, in both bulk GBM and GICs. These effects were linked to the DDR by their abrogation upon treatment with an ATM inhibitor [70]. In addition to supporting the maintenance and proliferation of the GIC phenotype, c-MET is also found to stimulate tumor angiogenesis by induction of vascular endothelial growth factor (VEGF) expression [80]. Furthermore, resistance to bevacizumab, an anti-VEGF monoclonal antibody, can occur through c-MET activation of pro-survival and invasion mechanisms [81]. Given the potential for tumor repopulation and recurrence afforded by processes stimulated by c-MET, the prospect of blocking ionizing radiation-induced c-MET signaling could have tremendous therapeutic benefit. Both in vitro and in vivo models have been used to test this hypothesis by targeting the c-MET receptor with genetic approaches in combination with ionizing radiation. Combined therapy with ionizing radiation and c-MET inhibition decreased cell proliferation and tumor growth compared to ionizing radiation or c-MET inhibition alone [80, 82], while another approach, targeting the c-MET ligand, HGF, with three neutralizing antibodies, also decreased tumor volume [83]. Most significantly, dual inhibition of the c-MET receptor and HGF-ligand expression combined with ionizing radiation reduced proliferation and tumor volume while increasing apoptosis, DNA fragmentation, and animal survival [84]. These data are strong support for investigating c-MET inhibitors such as cabozantinib (XL-184; Exelixis), in combination with conventional GBM therapy [85].

A number of clinical trials are currently assessing new drugs targeting HGF/c-MET signaling. Several of these drugs have completed studies in other solid tumors such as skin, lung, and thyroid cancers, all of which are often driven by similar molecular mechanisms found in GBM [86]. Cabozantinib, a pan-tyrosine kinase inhibitor with high affinity for c-MET and VEGFR2, is being testing in a phase II clinical trial for recurrent GBM [87, 88]. Most notably, cabozantinib is also currently under investigation in a phase I trial assessing it in combination with concurrent temozolomide and radiation therapy [85]. Another approach to therapy aimed at the HGF/c-MET pathway is ligand sequestration with a biologic drug. A monoclonal antibody against HGF, rilotumumab (AMG-102; Amgen), is currently under investigation in two phase II trials as a single-agent therapy for recurrent GBM and as a combination therapy with bevacizumab [89, 90].

Another component of radioresistance through the DDR may involve the Polycomb group protein BMI1. The Polycomb group proteins act as epigenetic silencers, and repress the expression of a range of proteins involved in the regulation of stem cell function during embryonic development and may be directly involved in tumor initiation [91]. BMI1 is part of the Polycomb repressive complex 1 (PRC1) and it has been found to be essential in the maintenance of the stem cell phenotype in both neural stem cells (NSCs) and GICs [92, 93]. Elevated expression of BMI1 in glioma correlates with poor patient survival [94]. It has been recently described that ionizing radiation stimulates the accumulation of BMI1 in chromatin and in DDR proteins. Knockdown of BMI1 impaired the DDR and increased GIC radiosensitivity [92]. While a mechanism of BMI1 in radioresistance is unclear, current evidence suggests that it represents a promising target for improving radiosensitization of GICs.

Thus far our discussion has focused on radioresistance mechanisms that relate directly to activation of the DDR in response to ionizing radiation, but GIC therapeutic resistance may involve molecular characteristics in addition to the radiation stress response. It is very likely that there are multiple, nonexclusive pathways that contribute to the radioresistance of GICs. It is also possible that pathways involved in radioresistance may overlap and have interplay with signaling involved in the maintenance and proliferation of the GIC phenotype, as evidenced by the previous discussion of the L1CAM surface marker. Two pieces of evidence support alternative pathways. First, preferential activation of CHK1 and CHK2 in GICs has been described in the absence of ionizing radiation [48, 49]. Second, enhanced GIC survival following ionizing radiation has been reported without differences in DNA repair capacity compared to non-GICs [95]. These findings suggest that elevated DNA repair in response to DDR activation may not be the only method of radiation resistance at work in GICs. A number of other signaling pathways that are important in the maintenance of the stem cell-like phenotype may play a role in the therapeutic resistance of GICs.

The NOTCH signaling pathway is a highly conserved regulator of cell fate in both embryonic and adult tissues. Its effect is largely dependent on the context of its stimulation, but in the majority of tissues it contributes to the maintenance of an undifferentiated state. Unsurprisingly, the NOTCH receptor is over-expressed in a variety of cancer stem cells including GICs [96, 97]. NOTCH is a cell surface receptor with an intracellular domain activated upon ligand binding similarly to L1CAM. Following the binding of its ligand, DELTA/JAGGED, the NOTCH receptor, is activated via proteolytic cleavage by γ-secretase to promote the release and nuclear translocation of the NOTCH intracellular domain (NICD) [98]. The activation of NICD promotes the activation of the PI3K/AKT pathway and expression of NOTCH-regulated genes. These genes include c-Myc, Hes1, and Hey1, which are responsible for promoting self-renewal and GIC maintenance [96, 99–101].

NOTCH is important for GICs in the absence of ionizing radiation. There is evidence that treatment with high concentrations of γ-secretase inhibitors decreases tumorsphere formation, proliferation, and xenograft growth, as well as increases differentiation. Ionizing radiation induces NOTCH activation in GICs, resulting in the expansion of the GIC population [100]. Inhibition of NOTCH similarly improves the radiosensitivity of GICs [102, 103]. γ-Secretase inhibitors also enhanced the radiation-induced cell death and impaired the clonogenic survival of GICs in comparison to non-GICs. Furthermore, knockdown of NOTCH sensitized GICs to radiation and impaired xenograft tumor growth. Exogenous expression of constitutively active NICD protected GICs from radiation and the effect of γ-secretase inhibitors was attenuated [100]. Importantly, the inhibition of NOTCH signaling did not demonstrate changes in the DDR of the GICs, but reduced the activity through the PI3K/AKT pathway in response to radiation therapy. Taken together this evidence supports the synergistic effect that γ-secretase inhibitors can have with radiation therapy in GBM treatment.

There are currently several clinical trials evaluating γ-secretase inhibitors in the treatment of patients with GBM [104]. One promising γ-secretase inhibitor is RO4929097 which has been studied in a phase I trial in combination with chemoradiotherapy for newly diagnosed glioma [105]. Investigation of this compound has also moved into phase II studies as a single agent in patients with recurrent GBM [106, 107]. RO4929097 is also being studied in phase II trials as combination therapy with the tyrosine kinase inhibitor cediranib (AZD2171/AstraZeneca) in multiple solid tumors, and with bevacizumab in patients with recurrent high-grade gliomas [108, 109].

The downstream mechanism of radiosensitization through NOTCH inhibition highlights an intracellular signaling axis that has long been studied in GBM biology, the PI3K/AKT axis [110]. This pathway is a mediator of cell survival and invasion signaling pathways and is commonly dysregulated in GBM [111, 112]. Similarly, upstream regulators of PI3K/AKT are found commonly mutated in GBM, increasing signaling through this axis [113–115]. This axis is known to be important for GIC maintenance, as direct inhibition of AKT alone preferentially increased apoptosis, and reduced neurosphere formation, migration, and invasion in GICs compared to non-GICs [116, 117]. In established GBM cell lines radiation therapy is found to stimulate AKT activation and increase both survival and invasion signaling [118–120]. Given the importance of PI3K/AKT in GIC biology it is reasonable to conclude that radiation therapy has similar effects on GICs, and a number of laboratories have found that inhibition of AKT improves radiosensitivity of both GICs and established GBM cell lines in vitro and in vivo [100, 121–124]. While different mechanisms of AKT-mediated radioresistance have been suggested, including effects on DNA repair capacity or the ability to overcome cell cycle arrest, the bottom line is that this pathway represents an integral target for therapeutic radiosensitization [125]. Further elucidation of the downstream effectors of PI3K/AKT involved in radioresistance will be important, but the number and variety of inhibitors of this pathway that are currently in clinical trials are promising.

Another cellular response pathway, downstream of PI3K/AKT, that is induced by radiation and may contribute to the radiation resistance of GICs is autophagy [126]. Autophagy is an intracellular degradation system that cells can use to break down and recycle their contents to provide an alternate source of energy in response to metabolic stress or starvation. This is an important homeostatic process which can contribute to therapeutic resistance in many cancers, or when unchecked can lead to cell death [126]. Ionizing radiation induces autophagy preferentially in GICs compared to non-GICs and GICs are found to express higher levels of autophagy-related proteins (LC3, ATG5, and ATG12) [127]. In further support for autophagy as a target for radiosensitization is evidence that autophagy inhibitors and gene silencing that targets autophagy genes reduce GIC survival and their ability to form neurospheres following radiation [127].

Unfortunately, the benefit of inhibiting autophagy in combination with radiation is not completely clear as other studies demonstrated that activation of autophagy instead of its inhibition can have a radiosensitizing effect. The mammalian target of rapamycin (mTOR) acts as a major checkpoint in the regulation of autophagy signaling, integrating stimulation via PI3K/AKT and the cell's nutrient sensing apparatus [128]. One approach to radiosensitization activated autophagy using inhibitors of the mTOR signaling pathway combined with radiation and observed an increase in radiosensitivity, neural differentiation, and a reduction in the self-renewal and proliferative capacities of GICs [129, 130]. Another approach, which used a combination of cilengitide, an α_v integrin inhibitor that is currently in clinical trials, and radiation to induce autophagy found that it enhanced cytotoxicity and decreased cell survival in GICs [131]. Taken together, the evidence supports a role for autophagy in the GIC response to ionizing radiation, but does not indicate a clear direction for therapeutic intervention.

The in vitro study of GICs has provided an excellent picture of intracellular mechanisms of GBM radioresistance, but some groups have questioned whether it is a sufficient model for studying the radioresistance that GBM displays in situ [132]. The relative difficulty in studying radiation survival curves (considered by many to be the gold standard in assessment of radiosensitivity) in GIC compared to non-GIC populations and conflicting evidence regarding the precise mechanisms of radioresistance (specifically preferential elevations in DSB repair capacity between GICs and non-GICs) suggest that in vitro analysis may be subject to cell line-dependent variability, and by itself is not optimal for therapeutic testing [47, 49]. Another approach, that has accumulated a strong body of evidence, is to assess the role of the tumor microenvironment in the radioresistance of GICs. This model is supported by evidence that stimuli found in the tumor microenvironment are integral in the maintenance and proliferation of GICs in vitro [133–136]. The microenvironment can contribute to differences in DNA repair capacity seen among different GIC lineages, as sections of irradiated tumors generated from GIC versus non-GIC cell populations display differences in DSB repair capacity while when the cell populations were irradiated in vitro the DSB repair capacities were similar [132]. This suggests that there are signaling cues found in vivo that drive the radioresistance of GICs.

The field of radiation biology has recognized for nearly 100 years that a cell's microenvironment can have protective or sensitizing effects on the DNA-damaging properties of ionizing radiation [137]. These effects can be physical, such as the availability of oxygen for the generation of DNA-damaging free radicals, or biological, such as molecular signaling that promotes DNA repair and cell proliferation. In GBM, the therapeutic implications of the microenvironment are becoming apparent as we gain insight into how GICs exist in the context of their tumoral location. The current understanding of GBM tumors identifies specific anatomical and functional locations within the tumor, termed "stem cell niches," where signaling cues and nutrient availability promote the survival and proliferation of GICs [136, 138]. GICs tend to cluster in niches characterized as perivascular and hypoxic, though there may be other general types.

The concept of a niche to support the stem cell phenotype is parallel to the microenvironments identified in the support and maintenance of neural stem cells (NSCs). The NSC niche, well characterized in murine models, is understood to be an interactive structural unit, concentrated around blood vessels, where the NSCs have access to signaling molecules, nutrition, and use of vasculature for migration [138]. GICs have similarly been found to be regulated by relationships to endothelial cells for their maintenance and self-renewal [139].

Paracrine signaling from endothelial cells can support GIC renewal and proliferation [140]. In an interesting in vitro model, GICs were cocultured with or without tumor microvascular endothelial cells (tMVEC) isolated from the same tumor specimen and exposed to radiation therapy and/or chemotherapy with temozolomide. GICs cultured with tMVECs not only recovered from the therapeutic insult more quickly, but the cultures were enriched for the GIC phenotype, as seen in recurrent GBM following chemoradiation [141]. This suggests that a combination of soluble and membrane-bound factors from endothelial cells can contribute to

GIC maintenance and radioresistance. Endothelial cell expression of the NOTCH ligand is found to drive GIC self-renewal and proliferation, in addition to the previously discussed role of NOTCH signaling in radioresistance [142, 143]. Furthermore, endothelial cell production of nitric oxide is found to stimulate transcription of NESTIN, a protein highly expressed in GICs, and Hes1, a target of NOTCH signaling. It was also found that the production of nitric oxide by endothelial cells stimulated the development of the stem cell phenotype in cultured GBM cell lines and the expansion of GICs [144].

GICs are not only passive receivers of this paracrine stimulus from endothelial cells. Lineage tracing studies of GICs in 21 GBM xenografts demonstrated that they gave rise to pericytes in vivo [145]. GICs are capable of recruiting endothelial cells and stimulating tube formation, supporting their active role in remodeling the microenvironment [146, 147]. GICs are key players in this dynamic process, giving rise to tumors with greater vascularity, necrosis, and hemorrhage compared to tumors generated from non-GICs [146]. The model for development of these qualities suggests that rapid growth of the tumor cells surpasses the supportive capacity of the available blood supply, creating hypoxic zones that stimulate angiogenic signals and give rise to disorganized tumor vasculature. Given that necrosis and angiogenesis are both characteristics of GBM, and that extensive necrosis is a negative prognostic factor in GBM patients, a number of therapies have attempted to target the molecular components that drive the development of these microenvironments [4]. One of the most promising of these therapies has been the monoclonal antibody against the cytokine vascular endothelial growth factor (VEGF), bevacizumab [148, 149].

GICs produce much higher levels of VEGF, upregulated 10–20-fold, compared to non-GICs, under both normoxia and hypoxia [146]. Both in vitro and in vivo evaluation of bevacizumab has shown that it is capable of abrogating the angiogenic signaling of GICs and reducing tumor vasculature [146]. Unfortunately, the use of bevacizumab clinically is presenting a more complicated picture, with recent evidence showing no survival benefit in the treatment of newly diagnosed GBM, but there may still be efficacy in progressive or recurrent disease [150]. VEGF is not the only important player in the signaling of the perivascular niche, and the relationship between GICs and endothelial cells has a number of implications for the use of ionizing radiation because interactive signaling with endothelial cells promotes GIC survival [139, 141].

GICs are recruited to endothelial cells via chemotactic signals with SDF-1/CXCR4 and were stimulated to differentiate into pericytes due in part to TGFβ secreted by endothelial cells. The cytokine SDF-1 plays a well-established role in the invasive behavior of GICs, in addition to exerting proliferative and antiapoptotic stimulus on a variety of glioma cell lines in vitro [151–153]. This axis may have a particularly detrimental role in GBM because its expression is induced by hypoxia and it can then support radioresistance and recurrence following ionizing radiation.

TGFβ is an intriguing cytokine because it is found to stimulate either tumor suppression or disease progression in different cell types and tumors [154]. In normal brain tissue TGFβ has an antiproliferative effect, whereas GBM tumors are known to express this cytokine abundantly and to proliferate in response to it [155]. Decades of research suggests that GBM tumors have overcome the antiproliferative effects, and instead the abundance of TGFβ produced by tumors may exert suppressive effects on the host antitumoral immune response [134, 156]. In GICs, TGFβ improves the tumorigenicity of injected cells in xenograft models and it stimulates transcription factors that play a role in stem cell maintenance [157–159].

Ionizing radiation induces the expression of TGFβ, likely through a mechanism that involves reactive oxygen species, and TGFβ has been directly linked to the DDR and radiosensitivity. In studies of a TGFβ inhibitor combined with ionizing radiation the neurosphere-forming capacity and repair of DNA damage were reduced in GICs and in bulk tumor specimens. There was a corresponding induction of self-renewal signals through NOTCH and CXCR4 when TGFβ inhibition and radiation therapy were combined suggesting a possible escape mechanism for radioresistance [160]. Taken together, this evidence supports a role for anti-TGFβ therapy in targeting GIC radioresistance.

Despite abundant angiogenic signaling in GBM, the rapid growth of the tumor cells will outstrip their ability to stimulate sufficient vessel growth. This phenomena is evident in the highly disorganized vessels and variable oxygen tension across GBM tumors [161]. Most solid tumors, including GBM, contain regions of irregular blood flow creating fluctuating and abnormal levels of oxygen tension [162]. Analysis of normal brain and glioma revealed that the physiological concentration of oxygen in healthy brain tissue ranges from 12.5 to 2.5 % (pO_2 = 100–20 mmHg). However in GBM masses there is mild to moderate/severe hypoxia with oxygen concentrations ranging between 2.5 and 0.5 % (pO_2 = 20–4 mmHg) for mild and 0.5–0.1 % (pO_2 = 4–0.75 mmHg) for severe hypoxia [163, 164]. The result of this loss of oxygen and nutrients is necrosis, a characteristic of GBM. Tumor hypoxia is a negative prognostic factor in GBM patients and is associated with tumor aggression. These correlations may be linked to GIC biology. The hypoxic niche paradoxically represents another supportive microenvironment for GICs as hypoxia increases the expression of some markers of GICs in glioma cells [165–167].

The necrotic cores of GBM have elevated expression of cellular markers of hypoxia and the GIC phenotype [168, 169]. Reduced oxygen levels are found to promote the formation of neurospheres in both GICs and non-GICs, and the stem cell genes *Sox2* and *Oct4* are upregulated in glioma cells under moderate hypoxia [170]. In both normal cells and tumors the response to hypoxia is mediated through induction of the hypoxia-inducible factors (HIFs) [162]. The HIF proteins are heterodimeric and exist as a beta subunit which is constitutively present in the nucleus, and alpha subunits which are typically cytosolic and degraded rapidly in the presence of oxygen. The HIF alpha subunits are analogous to an on/off switch, and when regulated by prolyl hydroxylase, which promotes their ubiquitination and degradation by the proteasome in well-oxygenated environments, they are off. In poorly oxygenated environments the function of prolyl hydroxylase is impaired, allowing the stabilization of the HIFα subunits, their translocation to the nucleus, binding of the beta subunit, and subsequent transcriptional activation of a number of target genes [171]. The alpha subunits HIF1α and HIF2α have well-characterized function, and while there is some overlap they differ in their activity at different levels of hypoxia and in their transcriptional targets [172, 173]. Both HIF1α and

HIF2α are critical for GIC function, with knockdown of either one individually reducing neurosphere formation of GICs in vitro, and in vivo their knockdown correlated with increased survival in mice bearing intracranial xenografts.

While both HIF1α and HIF2α are critical for GICs, HIF2α represents a more attractive therapeutic target. First, both neuronal stem cells and normal endothelial cells rely on HIF1α so targeting this protein will have a limited therapeutic index [174]. Second, HIF2α is preferentially expressed in GICs and upregulates the specific stem cell factors Nanog, Oct4, and Sox2 [30]. Uniquely, HIF2α can also promote the GIC phenotype in non-GIC cells [33, 170, 173]. Experimental expression of HIF2α in non-GICs induced the expression of the genes *Oct4*, *Nanog*, and *c-myc* [30]. Expression of a non-degradable HIF2α increased the ratio of GICs to non-GICs, and overexpression of HIF2α in non-GICs increased their tumorigenic potential in mouse xenograft models [173]. HIF proteins are also important mediators of angiogenic signals from GBM as VEGF is a downstream target of HIF as are other pro-angiogenic signals such as angiopoietins [162, 175, 176].

Taken together the data indicate that HIF2α targeting is an attractive approach to GICs and may be useful in combination with ionizing radiation as it could reduce the proportion of these highly resistant cells within a patient's tumor. The potential for broad spectrum drugs, such as the aminoglycoside digoxin, to decrease HIF protein levels in vitro and inhibit tumor growth in xenografts has been demonstrated [177]. However, this therapy would suffer from the therapeutic index limitations due to its targeting of both HIF1α and HIF2α. It is possible to develop HIF2α specific inhibitors, but unfortunately there are no drugs currently under clinical evaluation [178].

Beyond contributing to the maintenance and potential expansion of the GIC population, tissue hypoxia represents a challenge to radiation therapy in a mechanistic way. A major concept in radiobiology, for more than 60 years, is the recognition that the proportion of hypoxic cells in a tissue decreases its radioresponsiveness. In tumors, the hypoxic cells tend to be clustered in the center of the mass, most distant from the vasculature. The challenge posed by the hypoxic cores of most solid tumors is that the cells can still give rise to recurrence. The hypoxic cores of most solid tumors display radioresistance relative to their level of hypoxia, with a dose modifying effect. The dose modifying effect of the hypoxia found commonly in GBM tumors is such that to achieve the equivalent cytotoxic effect desired at normoxia, three-times the radiation dose must be administered. This suggests that an improvement in tumor hypoxia could dramatically improve the effect of ionizing radiation in GBM.

Because radiation therapy is a mainstay in the treatment of GBM its effect on the normal tissue of the brain, in addition to the tumor, are necessary to assess its therapeutic index. It is also important to understand the impact of radiation on the microenvironments and cellular changes induced in the brain of a GBM patient. Historically, this has not been a major focus of research in the GBM field, but as we learn more about extensive changes in gene expression in both normal and cancer cells induced by radiation we need to give more consideration to these effects [179–181]. One population of normal cells that are particularly sensitive to the effects of radiation are the NSCs. The relationship between GICs and NSCs is an area of intense study, and the striking similarities in phenotypes and overlapping signaling systems that support each population's maintenance are important to recognize. The subgranular zone (SGZ) of the hippocampus and the subventricular zone (SVZ) of the lateral periventricular region have been found to harbor and maintain a population of NSCs that can differentiate, migrate, and integrate into other functional brain regions [182].

NSCs are exquisitely radiosensitive, and the regenerative capacity of the cells in both the SGZ and SVZ can be impaired by even low to moderate levels (0.5–5 Gy) of radiation, and these changes may persist for as long as 25 months in rodents and humans [183–185]. While there is significant evidence linking damage (either cytotoxic chemotherapy or radiation induced) to the SGZ and hippocampus to neurocognitive decline, there is less evidence linking damage to the SVZ and decline in function [186]. A better understanding of the potential damage that radiation to the SVZ may cause patients will be necessary in the future.

Similarly important is the need to address questions regarding interaction between NSCs and GBM. Given that certain autocrine/paracrine signaling loops are implicated in the maintenance and survival of both NSCs and GICs, it is possible that NSC niches could support GIC development or even provide a pool for their population to arise from. There is currently no direct evidence elucidating the relationship between NSCs and GICs in tumor cells; however, preclinical models support a number of hypotheses that link the SVZ to tumor recurrence. Tumor-suppressor gene deletions in NSCs have shown that they can be a source of tumorigenesis, giving rise to tumors in the brain that resemble the invasive and malignant potential of human gliomas [187–191]. Models in mice and rats that investigated the vascular niche thought to support the NSCs of the SVZ have shown the potential for secreted factors in this microenvironment to induce glioma-like hyperplasias [192], and migration patterns of NSCs have been found to infiltrate gliomas with both supportive and inhibitory effects on glioma progression [193, 194].

While the nature of the relationship between GICs and NSC niches is currently unclear, a number of studies have implicated the SVZ as a target of radiation to sterilize possible microinvasive disease [195]. Clinical evidence similarly suggests a relationship between glioma progression and involvement of the SVZ. Patients that have tumors contacting the SVZ have been found to have a poorer survival compared to those that do not contact the SVZ [196, 197]. Furthermore, retrospective clinical evaluations are interrogating the role of radiation in a possible therapeutic enhancement targeting the SVZ [198, 199].

In conclusion, the study of GICs has provided us with a powerful model to understand the radioresistance of GBM. The heterogeneity inherent in GBM is partially responsible for the difficulty in finding effective treatment. It is clear that GICs possess a variety of intracellular adaptations allowing them to preferentially survive and even proliferate in response to radiation. This model has helped to identify or support a number of potential targets for therapeutic radiosensitization, which are under clinical investigation to improve patient survival. Unfortunately such targeted therapies may fall short if factors present in the microenvironment of this disease persist or are even recapitulated by our treatments. The possibility that cellular phenotypes within the tumor are dynamically regulated by extrinsic stress and signaling, such as that stimulated by radiation, suggests that modulation of radiation's effectiveness alone may not be enough to improve patient survival. Further investigation of radiation therapy in GBM and other stem cell neoplasms of the central nervous system should focus on disrupting the intracellular mechanisms of resistance and microenvironmental stimuli.

References

1. Dolecek TA, Propp JM, Stroup NE, Kruchko C. CBTRUS statistical report: primary brain and central nervous system tumors diagnosed in the United States in 2005-2009. Neuro Oncol. 2012;14 Suppl 5:v1–49.
2. Ostrom QT, Gittleman H, Farah P, Ondracek A, Chen Y, Wolinsky Y, Stroup NE, Kruchko C, Barnholtz-Sloan JS. CBTRUS statistical report: primary brain and central nervous system tumors diagnosed in the United States in 2006–2010. Neuro Oncol. 2013;15 Suppl 2:ii1–56.
3. Stupp R, Mason WP, van den Bent MJ, Weller M, Fisher B, Taphoorn MJB, Belanger K, Brandes AA, Marosi C, Bogdahn U, Curschmann J, Janzer RC, Ludwin SK, Gorlia T, Allgeier A, Lacombe D, Cairncross JG, Eisenhauer E, Mirimanoff RO. Radiotherapy plus concomitant and adjuvant temozolomide for glioblastoma. N Engl J Med. 2005;352(10): 987–96.
4. Lacroix M, Abi-Said D, Fourney DR, Gokaslan ZL, Shi W, DeMonte F, Lang FF, McCutcheon IE, Hassenbusch SJ, Holland E, Hess K, Michael C, Miller D, Sawaya R. A multivariate analysis of 416 patients with glioblastoma multiforme: prognosis, extent of resection, and survival. J Neurosurg. 2001;95(2):190–8.
5. Walker MD, Strike TA, Sheline GE. An analysis of dose-effect relationship in the radiotherapy of malignant gliomas. Int J Radiat Oncol Biol Phys. 1979;5(10):1725–31.
6. Bleehen NM, Stenning SP. A Medical Research Council trial of two radiotherapy doses in the treatment of grades 3 and 4 astrocytoma. The Medical Research Council Brain Tumour Working Party. Br J Cancer. 1991;64(4):769–74.
7. Chaichana KL, Cabrera-Aldana EE, Jusue-Torres I, Wijesekera O, Olivi A, Rahman M, Quinones-Hinojosa A. When gross total resection of a glioblastoma is possible, how much resection should be achieved? World Neurosurg. 2014;82(1–2):e257–65.
8. Shapiro WR, Young DF. Treatment of malignant glioma. A controlled study of chemotherapy and irradiation. Arch Neurol. 1976;33(7):494–550.
9. Walker MD, Alexander Jr E, Hunt WE, MacCarty CS, Mahaley Jr MS, Mealey Jr J, Norrell HA, Owens G, Ransohoff J, Wilson CB, Gehan EA, Strike TA. Evaluation of BCNU and/or radiotherapy in the treatment of anaplastic gliomas. A cooperative clinical trial. J Neurosurg. 1978;49(3):333–43.
10. Lee SW, Fraass BA, Marsh LH, Herbort K, Gebarski SS, Martel MK, Radany EH, Lichter AS, Sandler HM. Patterns of failure following high-dose 3-D conformal radiotherapy for high-grade astrocytomas: a quantitative dosimetric study. Int J Radiat Oncol Biol Phys. 1999;43(1):79–88.
11. Walker MD, Green SB, Byar DP, Alexander Jr E, Batzdorf U, Brooks WH, Hunt WE, MacCarty CS, Mahaley Jr MS, Mealey Jr J, Owens G, Ransohoff 2nd J, Robertson JT, Shapiro WR, Smith Jr KR, Wilson CB, Strike TA. Randomized comparisons of radiotherapy and nitrosoureas for the treatment of malignant glioma after surgery. N Engl J Med. 1980;303(23):1323–9.
12. Shapiro WR, Green SB, Burger PC, Mahaley Jr MS, Selker RG, VanGilder JC, Robertson JT, Ransohoff J, Mealey Jr J, Strike TA. Randomized trial of three chemotherapy regimens and

two radiotherapy regimens and two radiotherapy regimens in postoperative treatment of malignant glioma. Brain Tumor Cooperative Group Trial 8001. J Neurosurg. 1989; 71(1):1–9.
13. DeAngelis LM, Delattre JY, Posner JB. Radiation-induced dementia in patients cured of brain metastases. Neurology. 1989;39(6):789–96.
14. Chang CH, Horton J, Schoenfeld D, Salazer O, Perez-Tamayo R, Kramer S, Weinstein A, Nelson JS, Tsukada Y. Comparison of postoperative radiotherapy and combined postoperative radiotherapy and chemotherapy in the multidisciplinary management of malignant gliomas. A joint Radiation Therapy Oncology Group and Eastern Cooperative Oncology Group study. Cancer. 1983;52(6):997–1007.
15. Nelson DF, Diener-West M, Horton J, Chang CH, Schoenfeld D, Nelson JS. Combined modality approach to treatment of malignant gliomas–re-evaluation of RTOG 7401/ECOG 1374 with long-term follow-up: a joint study of the Radiation Therapy Oncology Group and the Eastern Cooperative Oncology Group. NCI Monogr. 1988;(6):279–84.
16. Hochberg FH, Pruitt A. Assumptions in the radiotherapy of glioblastoma. Neurology. 1980;30(9):907–11.
17. Wallner KE, Galicich JH, Krol G, Arbit E, Malkin MG. Patterns of failure following treatment for glioblastoma multiforme and anaplastic astrocytoma. Int J Radiat Oncol Biol Phys. 1989;16(6):1405–9.
18. Thornton Jr AF, Sandler HM, Ten Haken RK, McShan DL, Fraass BA, La Vigne ML, Yanke BR. The clinical utility of magnetic resonance imaging in 3-dimensional treatment planning of brain neoplasms. Int J Radiat Oncol Biol Phys. 1992;24(4):767–75.
19. Lattanzi JP, Fein DA, McNeeley SW, Shaer AH, Movsas B, Hanks GE. Computed tomography-magnetic resonance image fusion: a clinical evaluation of an innovative approach for improved tumor localization in primary central nervous system lesions. Radiat Oncol Investig. 1997;5(4):195–205.
20. Fiorentino A, Caivano R, Pedicini P, Fusco V. Clinical target volume definition for glioblastoma radiotherapy planning: magnetic resonance imaging and computed tomography. Clin Transl Oncol. 2013;15(9):754–8.
21. Pirzkall A, McKnight TR, Graves EE, Carol MP, Sneed PK, Wara WW, Nelson SJ, Verhey LJ, Larson DA. MR-spectroscopy guided target delineation for high-grade gliomas. Int J Radiat Oncol Biol Phys. 2001;50(4):915–28.
22. Petrecca K, Guiot M-C, Panet-Raymond V, Souhami L. Failure pattern following complete resection plus radiotherapy and temozolomide is at the resection margin in patients with glioblastoma. J Neurooncol. 2013;111(1):19–23.
23. Rahman M, Deleyrolle L, Vedam-Mai V, Azari H, Abd-El-Barr M, Reynolds BA. The cancer stem cell hypothesis: failures and pitfalls. Neurosurgery. 2011;68(2):531–45. discussion 545.
24. Reya T, Morrison SJ, Clarke MF, Weissman IL. Stem cells, cancer, and cancer stem cells. Nature. 2001;414(6859):105–11.
25. Zhou B-BS, Zhang H, Damelin M, Geles KG, Grindley JC, Dirks PB. Tumour-initiating cells: challenges and opportunities for anticancer drug discovery. Nat Rev Drug Discov. 2009;8(10):806–23.
26. O'Connor ML, Xiang D, Shigdar S, Macdonald J, Li Y, Wang T, Pu C, Wang Z, Qiao L, Duan W. Cancer stem cells: a contentious hypothesis now moving forward. Cancer Lett. 2014; 344(2):180–7.
27. Lathia JD, Gallagher J, Myers JT, Li M, Vasanji A, McLendon RE, Hjelmeland AB, Huang AY, Rich JN. Direct in vivo evidence for tumor propagation by glioblastoma cancer stem cells. PLoS One. 2011;6(9):e24807.
28. Singh SK, Hawkins C, Clarke ID, Squire JA, Bayani J, Hide T, Henkelman RM, Cusimano MD, Dirks PB. Identification of human brain tumour initiating cells. Nature. 2004;432(7015):396–401.
29. Schonberg DL, Lubelski D, Miller TE, Rich JN. Brain tumor stem cells: molecular characteristics and their impact on therapy. Mol Aspects Med. 2014;39:82–101.

30. Heddleston JM, Li Z, McLendon RE, Hjelmeland AB, Rich JN. The hypoxic microenvironment maintains glioblastoma stem cells and promotes reprogramming towards a cancer stem cell phenotype. Cell Cycle. 2009;8(20):3274–84.
31. Hjelmeland AB, Wu Q, Heddleston JM, Choudhary GS, MacSwords J, Lathia JD, McLendon R, Lindner D, Sloan A, Rich JN. Acidic stress promotes a glioma stem cell phenotype. Cell Death Differ. 2011;18(5):829–40.
32. Soeda A, Park M, Lee D, Mintz A, Androutsellis-Theotokis A, McKay RD, Engh J, Iwama T, Kunisada T, Kassam AB, Pollack IF, Park DM. Hypoxia promotes expansion of the CD133-positive glioma stem cells through activation of HIF-1alpha. Oncogene. 2009;28(45): 3949–59.
33. Seidel S, Garvalov BK, Wirta V, von Stechow L, Schänzer A, Meletis K, Wolter M, Sommerlad D, Henze A-T, Nistér M, Reifenberger G, Lundeberg J, Frisén J, Acker T. A hypoxic niche regulates glioblastoma stem cells through hypoxia inducible factor 2 alpha. Brain J Neurol. 2010;133(Pt 4):983–95.
34. Cheng L, Bao S, Rich JN. Potential therapeutic implications of cancer stem cells in glioblastoma. Biochem Pharmacol. 2010;80(5):654–65.
35. Roos WP, Kaina B. DNA damage-induced apoptosis: From specific DNA lesions to the DNA damage response and apoptosis. Cancer Lett. 2012;332(2):237–48.
36. Ciccia A, Elledge SJ. The DNA damage response: making it safe to play with knives. Mol Cell. 2010;40(2):179–204.
37. Lord CJ, Ashworth A. The DNA damage response and cancer therapy. Nature. 2012;481(7381):287–94.
38. Bartkova J, Hamerlik P, Stockhausen M-T, Ehrmann J, Hlobilkova A, Laursen H, Kalita O, Kolar Z, Poulsen HS, Broholm H, Lukas J, Bartek J. Replication stress and oxidative damage contribute to aberrant constitutive activation of DNA damage signalling in human gliomas. Oncogene. 2010;29(36):5095–102.
39. Bao S, Wu Q, McLendon RE, Hao Y, Shi Q, Hjelmeland AB, Dewhirst MW, Bigner DD, Rich JN. Glioma stem cells promote radioresistance by preferential activation of the DNA damage response. Nature. 2006;444(7120):756–60.
40. Firat E, Gaedicke S, Tsurumi C, Esser N, Weyerbrock A, Niedermann G. Delayed cell death associated with mitotic catastrophe in γ-irradiated stem-like glioma cells. Radiat Oncol. 2011;6:71.
41. Tamura K, Aoyagi M, Ando N, Ogishima T, Wakimoto H, Yamamoto M, Ohno K. Expansion of CD133-positive glioma cells in recurrent de novo glioblastomas after radiotherapy and chemotherapy. J Neurosurg. 2013;119(5):1145–55.
42. Chan JL, Lee SW, Fraass BA, Normolle DP, Greenberg HS, Junck LR, Gebarski SS, Sandler HM. Survival and failure patterns of high-grade gliomas after three-dimensional conformal radiotherapy. J Clin Oncol. 2002;20(6):1635–42.
43. Souhami L, Seiferheld W, Brachman D, Podgorsak EB, Werner-Wasik M, Lustig R, Schultz CJ, Sause W, Okunieff P, Buckner J, Zamorano L, Mehta MP, Curran Jr WJ. Randomized comparison of stereotactic radiosurgery followed by conventional radiotherapy with carmustine to conventional radiotherapy with carmustine for patients with glioblastoma multiforme: report of Radiation Therapy Oncology Group 93-05 protocol. Int J Radiat Oncol Biol Phys. 2004;60(3):853–60.
44. Tsao MN, Mehta MP, Whelan TJ, Morris DE, Hayman JA, Flickinger JC, Mills M, Rogers CL, Souhami L. The American Society for Therapeutic Radiology and Oncology (ASTRO) evidence-based review of the role of radiosurgery for malignant glioma. Int J Radiat Oncol Biol Phys. 2005;63(1):47–55.
45. Laperriere NJ, Leung PM, McKenzie S, Milosevic M, Wong S, Glen J, Pintilie M, Bernstein M. Randomized study of brachytherapy in the initial management of patients with malignant astrocytoma. Int J Radiat Oncol Biol Phys. 1998;41(5):1005–11.
46. Selker RG, Shapiro WR, Burger P, Blackwood MS, Arena VC, Gilder JC, Malkin MG, Mealey Jr JJ, Neal JH, Olson J, Robertson JT, Barnett GH, Bloomfield S, Albright R,

Hochberg FH, Hiesiger E, Green S, Brain Tumor Cooperative Group. The Brain Tumor Cooperative Group NIH Trial 87-01: a randomized comparison of surgery, external radiotherapy, and carmustine versus surgery, interstitial radiotherapy boost, external radiation therapy, and carmustine. Neurosurgery. 2002;51(2):343–55. discussion 355–7.
47. McCord AM, Jamal M, Williams ES, Camphausen K, Tofilon PJ. CD133+ glioblastoma stem-like cells are radiosensitive with a defective DNA damage response compared with established cell lines. Clin Cancer Res. 2009;15(16):5145–53.
48. Squatrito M, Brennan CW, Helmy K, Huse JT, Petrini JH, Holland EC. Loss of ATM/Chk2/p53 pathway components accelerates tumor development and contributes to radiation resistance in gliomas. Cancer Cell. 2010;18(6):619–29.
49. Ropolo M, Daga A, Griffero F, Foresta M, Casartelli G, Zunino A, Poggi A, Cappelli E, Zona G, Spaziante R, Corte G, Frosina G. Comparative analysis of DNA repair in stem and nonstem glioma cell cultures. Mol Cancer Res. 2009;7(3):383–92.
50. Williams RS, Williams JS, Tainer JA. Mre11-Rad50-Nbs1 is a keystone complex connecting DNA repair machinery, double-strand break signaling, and the chromatin template. Biochem Cell Biol Biochim Biol Cell. 2007;85(4):509–20.
51. Maser RS, Monsen KJ, Nelms BE, Petrini JH. hMre11 and hRad50 nuclear foci are induced during the normal cellular response to DNA double-strand breaks. Mol Cell Biol. 1997; 17(10):6087–96.
52. Lavin MF. ATM and the Mre11 complex combine to recognize and signal DNA double-strand breaks. Oncogene. 2007;26(56):7749–58.
53. Matsuoka S, Ballif BA, Smogorzewska A, McDonald 3rd ER, Hurov KE, Luo J, Bakalarski CE, Zhao Z, Solimini N, Lerenthal Y, Shiloh Y, Gygi SP, Elledge SJ. ATM and ATR substrate analysis reveals extensive protein networks responsive to DNA damage. Science. 2007; 316(5828):1160–6.
54. Bao S, Wu Q, Li Z, Sathornsumetee S, Wang H, McLendon RE, Hjelmeland AB, Rich JN. Targeting cancer stem cells through L1CAM suppresses glioma growth. Cancer Res. 2008;68(15):6043–8.
55. Cheng L, Wu Q, Huang Z, Guryanova OA, Huang Q, Shou W, Rich JN, Bao S. L1CAM regulates DNA damage checkpoint response of glioblastoma stem cells through NBS1. EMBO J. 2011;30(5):800–13.
56. Chiang Y-C, Teng S-C, Su Y-N, Hsieh F-J, Wu K-J. c-Myc directly regulates the transcription of the NBS1 gene involved in DNA double-strand break repair. J Biol Chem. 2003;278(21):19286–91.
57. Guerra L, Albihn A, Tronnersjö S, Yan Q, Guidi R, Stenerlöw B, Sterzenbach T, Josenhans C, Fox JG, Schauer DB, Thelestam M, Larsson L-G, Henriksson M, Frisan T. Myc is required for activation of the ATM-dependent checkpoints in response to DNA damage. PloS One. 2010;5(1):e8924.
58. Benetatos L, Vartholomatos G, Hatzimichael E. Polycomb group proteins and MYC: the cancer connection. Cell Mol Life Sci. 2014;71(2):257–69.
59. Vita M, Henriksson M. The Myc oncoprotein as a therapeutic target for human cancer. Semin Cancer Biol. 2006;16(4):318–30.
60. Wang J, Wang H, Li Z, Wu Q, Lathia JD, McLendon RE, Hjelmeland AB, Rich JN. c-Myc is required for maintenance of glioma cancer stem cells. PloS One. 2008;3(11):e3769.
61. Lee J-H, Paull TT. Direct activation of the ATM protein kinase by the Mre11/Rad50/Nbs1 complex. Science. 2004;304(5667):93–6.
62. Lee J-H, Paull TT. ATM activation by DNA double-strand breaks through the Mre11-Rad50-Nbs1 complex. Science. 2005;308(5721):551–4.
63. Matsuoka S, Huang M, Elledge SJ. Linkage of ATM to cell cycle regulation by the Chk2 protein kinase. Science. 1998;282(5395):1893–7.
64. Guo Z, Kumagai A, Wang SX, Dunphy WG. Requirement for Atr in phosphorylation of Chk1 and cell cycle regulation in response to DNA replication blocks and UV-damaged DNA in Xenopus egg extracts. Genes Dev. 2000;14(21):2745–56.

65. Gobbini E, Cesena D, Galbiati A, Lockhart A, Longhese MP. Interplays between ATM/Tel1 and ATR/Mec1 in sensing and signaling DNA double-strand breaks. DNA Repair. 2013; 12(10):791–9.
66. Zhou BB, Elledge SJ. The DNA damage response: putting checkpoints in perspective. Nature. 2000;408(6811):433–9.
67. Tribius S, Pidel A, Casper D. ATM protein expression correlates with radioresistance in primary glioblastoma cells in culture. Int J Radiat Oncol Biol Phys. 2001;50(2):511–23.
68. Nadkarni A, Shrivastav M, Mladek AC, Schwingler PM, Grogan PT, Chen J, Sarkaria JN. ATM inhibitor KU-55933 increases the TMZ responsiveness of only inherently TMZ sensitive GBM cells. J Neurooncol. 2012;110(3):349–57.
69. Vecchio D, Daga A, Carra E, Marubbi D, Baio G, Neumaier CE, Vagge S, Corvò R, Pia Brisigotti M, Louis Ravetti J, Zunino A, Poggi A, Mascelli S, Raso A, Frosina G. Predictability, efficacy and safety of radiosensitization of glioblastoma-initiating cells by the ATM inhibitor KU-60019. Int J Cancer. 2014;135(2):479–91.
70. De Bacco F, Luraghi P, Medico E, Reato G, Girolami F, Perera T, Gabriele P, Comoglio PM, Boccaccio C. Induction of MET by ionizing radiation and its role in radioresistance and invasive growth of cancer. J Natl Cancer Inst. 2011;103(8):645–61.
71. Organ SL, Tsao M-S. An overview of the c-MET signaling pathway. Ther Adv Med Oncol. 2011;3(1 Suppl):S7–19.
72. Cancer Genome Atlas Research Network. Comprehensive genomic characterization defines human glioblastoma genes and core pathways. Nature. 2008;455(7216):1061–8.
73. Snuderl M, Fazlollahi L, Le LP, Nitta M, Zhelyazkova BH, Davidson CJ, Akhavanfard S, Cahill DP, Aldape KD, Betensky RA, Louis DN, Iafrate AJ. Mosaic amplification of multiple receptor tyrosine kinase genes in glioblastoma. Cancer Cell. 2011;20(6):810–7.
74. Kong D-S, Song S-Y, Kim D-H, Joo KM, Yoo J-S, Koh JS, Dong SM, Suh Y-L, Lee J-I, Park K, Kim JH, Nam D-H. Prognostic significance of c-Met expression in glioblastomas. Cancer. 2009;115(1):140–8.
75. Brennan C, Momota H, Hambardzumyan D, Ozawa T, Tandon A, Pedraza A, Holland E. Glioblastoma subclasses can be defined by activity among signal transduction pathways and associated genomic alterations. PloS One. 2009;4(11):e7752.
76. Maher EA, Brennan C, Wen PY, Durso L, Ligon KL, Richardson A, Khatry D, Feng B, Sinha R, Louis DN, Quackenbush J, Black PM, Chin L, DePinho RA. Marked genomic differences characterize primary and secondary glioblastoma subtypes and identify two distinct molecular and clinical secondary glioblastoma entities. Cancer Res. 2006;66(23):11502–13.
77. Dunn GP, Rinne ML, Wykosky J, Genovese G, Quayle SN, Dunn IF, Agarwalla PK, Chheda MG, Campos B, Wang A, Brennan C, Ligon KL, Furnari F, Cavenee WK, Depinho RA, Chin L, Hahn WC. Emerging insights into the molecular and cellular basis of glioblastoma. Genes Dev. 2012;26(8):756–84.
78. Joo KM, Jin J, Kim E, Ho Kim K, Kim Y, Kang BG, Kang Y-J, Lathia JD, Cheong KH, Song PH, Kim H, Seol HJ, Kong D-S, Lee J-I, Rich JN, Lee J, Nam D-H. MET signaling regulates glioblastoma stem cells. Cancer Res. 2012;72(15):3828–38.
79. Vogel S, Peters C, Etminan N, Börger V, Schimanski A, Sabel MC, Sorg RV. Migration of mesenchymal stem cells towards glioblastoma cells depends on hepatocyte-growth factor and is enhanced by aminolaevulinic acid-mediated photodynamic treatment. Biochem Biophys Res Commun. 2013;431(3):428–32.
80. Abounader R, Ranganathan S, Lal B, Fielding K, Book A, Dietz H, Burger P, Laterra J. Reversion of human glioblastoma malignancy by U1 small nuclear RNA/ribozyme targeting of scatter factor/hepatocyte growth factor and c-met expression. J Natl Cancer Inst. 1999;91(18):1548–56.
81. Lu KV, Chang JP, Parachoniak CA, Pandika MM, Aghi MK, Meyronet D, Isachenko N, Fouse SD, Phillips JJ, Cheresh DA, Park M, Bergers G. VEGF inhibits tumor cell invasion and mesenchymal transition through a MET/VEGFR2 complex. Cancer Cell. 2012;22(1):21–35.

82. Jin J, Bae KH, Yang H, Lee SJ, Kim H, Kim Y, Joo KM, Seo SW, Park TG, Nam D-H. In vivo specific delivery of c-Met siRNA to glioblastoma using cationic solid lipid nanoparticles. Bioconjug Chem. 2011;22(12):2568–72.
83. Cao B, Su Y, Oskarsson M, Zhao P, Kort EJ, Fisher RJ, Wang LM, Vande Woude GF. Neutralizing monoclonal antibodies to hepatocyte growth factor/scatter factor (HGF/SF) display antitumor activity in animal models. Proc Natl Acad Sci U S A. 2001; 98(13):7443–8.
84. Lal B, Xia S, Abounader R, Laterra J. Targeting the c-Met pathway potentiates glioblastoma responses to gamma-radiation. Clin Cancer Res. 2005;11(12):4479–86.
85. Exelixis. Safety study of XL184 (Cabozantinib) in combination with temozolomide and radiation therapy in the initial treatment of adults with glioblastoma. [Online]. Available http://clinicaltrials.gov/ct2/show/NCT00960492?term=cabozantinib&rank=10. Accessed 11 Mar 2014.
86. Liu X, Newton RC, Scherle PA. Developing c-MET pathway inhibitors for cancer therapy: progress and challenges. Trends Mol Med. 2010;16(1):37–45.
87. Zhang Y, Guessous F, Kofman A, Schiff D, Abounader R. XL-184, a MET, VEGFR-2 and RET kinase inhibitor for the treatment of thyroid cancer, glioblastoma multiforme and NSCLC. IDrugs Investig Drugs J. 2010;13(2):112–21.
88. Exelixis. Study of XL184 (Cabozantinib) in adults with glioblastoma multiforme. [Online]. Available http://clinicaltrials.gov/ct2/show/NCT00704288?term=cabozantinib&rank=43. Accessed 11 Mar 2014.
89. Amgen. A phase II study to treat advanced malignant glioma.[Online]. Available http://www.clinicaltrials.gov/ct2/show/NCT00427440?term=rilotumumab&rank=12. Accessed 11 Mar 2014.
90. Peters K, Amgen. AMG 102 and avastin for recurrent malignant glioma. [Online]. Available http://www.clinicaltrials.gov/ct2/show/NCT01113398?term=rilotumumab&rank=6. Accessed 11 Mar 2014.
91. Acquati S, Greco A, Licastro D, Bhagat H, Ceric D, Rossini Z, Grieve J, Shaked-Rabi M, Henriquez NV, Brandner S, Stupka E, Marino S. Epigenetic regulation of survivin by Bmi1 is cell type specific during corticogenesis and in gliomas. Stem Cells (Dayton Ohio). 2013;31(1):190–202.
92. Facchino S, Abdouh M, Chatoo W, Bernier G. BMI1 confers radioresistance to normal and cancerous neural stem cells through recruitment of the DNA damage response machinery. J Neurosci. 2010;30(30):10096–111.
93. Valk-Lingbeek ME, Bruggeman SWM, van Lohuizen M. Stem cells and cancer; the polycomb connection. Cell. 2004;118(4):409–18.
94. Häyry V, Tynninen O, Haapasalo HK, Wölfer J, Paulus W, Hasselblatt M, Sariola H, Paetau A, Sarna S, Niemelä M, Wartiovaara K, Nupponen NN. Stem cell protein BMI-1 is an independent marker for poor prognosis in oligodendroglial tumours. Neuropathol Appl Neurobiol. 2008;34(5):555–63.
95. Jamal M, Rath BH, Williams ES, Camphausen K, Tofilon PJ. Microenvironmental regulation of glioblastoma radioresponse. Clin Cancer Res. 2010;16(24):6049–59.
96. Wang J, Sullenger BA, Rich JN. Notch signaling in cancer stem cells. Adv Exp Med Biol. 2012;727:174–85.
97. Rizzo P, Osipo C, Foreman K, Golde T, Osborne B, Miele L. Rational targeting of Notch signaling in cancer. Oncogene. 2008;27(38):5124–31.
98. Guruharsha KG, Kankel MW, Artavanis-Tsakonas S. The Notch signalling system: recent insights into the complexity of a conserved pathway. Nat Rev Genet. 2012;13(9):654–66.
99. Stockhausen M-T, Kristoffersen K, Poulsen HS. The functional role of Notch signaling in human gliomas. Neuro Oncol. 2010;12(2):199–211.
100. Wang J, Wakeman TP, Lathia JD, Hjelmeland AB, Wang X-F, White RR, Rich JN, Sullenger BA. Notch promotes radioresistance of glioma stem cells. Stem Cells (Dayton Ohio). 2010;28(1):17–28.

101. Barcellos-Hoff MH. Radiation-induced transforming growth factor beta and subsequent extracellular matrix reorganization in murine mammary gland. Cancer Res. 1993;53(17): 3880–6.
102. Fan X, Khaki L, Zhu TS, Soules ME, Talsma CE, Gul N, Koh C, Zhang J, Li Y-M, Maciaczyk J, Nikkhah G, Dimeco F, Piccirillo S, Vescovi AL, Eberhart CG. NOTCH pathway blockade depletes CD133-positive glioblastoma cells and inhibits growth of tumor neurospheres and xenografts. Stem Cells (Dayton Ohio). 2010;28(1):5–16.
103. Hovinga KE, Shimizu F, Wang R, Panagiotakos G, Van Der Heijden M, Moayedpardazi H, Correia AS, Soulet D, Major T, Menon J, Tabar V. Inhibition of notch signaling in glioblastoma targets cancer stem cells via an endothelial cell intermediate. Stem Cells (Dayton Ohio). 2010;28(6):1019–29.
104. Rivera M, Sukhdeo K, Yu J. Ionizing radiation in glioblastoma initiating cells. Front Oncol. 2013;3:74.
105. National Cancer Institute. RO4929097, temozolomide, and radiation therapy in treating patients with newly diagnosed malignant glioma. [Online]. http://www.clinicaltrials.gov/ct2/show/NCT01119599?term=RO4929097&rank=25. [Accessed: 11-Mar-2014].
106. National Cancer Institute. Gamma-secretase/notch signalling pathway inhibitor ro4929097 in treating patients with recurrent or progressive glioblastoma. [Online]. Available: http://www.clinicaltrials.gov/ct2/show/NCT01122901?term=RO4929097&rank=6. Accessed 11 Mar 2014.
107. National Cancer Institute. RO4929097 in treating patients with recurrent invasive gliomas. [Online]. Available http://www.clinicaltrials.gov/ct2/show/NCT01269411?term=RO4929097&rank=18. Accessed 11 Mar 2014.
108. National Cancer Institute. Gamma-secretase inhibitor RO4929097 and Cediranib Maleate in Treating Patients With Advanced Solid Tumors. [Online]. Available http://www.clinicaltrials.gov/ct2/show/NCT01131234?term=RO4929097&rank=13. Accessed 11 Mar 2014.
109. National Cancer Institute. RO4929097 and Bevacizumab in Treating Patients With Progressive or Recurrent Malignant Glioma. [Online]. Available http://www.clinicaltrials.gov/ct2/show/NCT01189240?term=RO4929097&rank=31. Accessed: 11 Mar 2014.
110. Hambardzumyan D, Becher OJ, Holland EC. Cancer stem cells and survival pathways. Cell Cycle. 2008;7(10):1371–8.
111. Broderick DK, Di C, Parrett TJ, Samuels YR, Cummins JM, McLendon RE, Fults DW, Velculescu VE, Bigner DD, Yan H. Mutations of PIK3CA in anaplastic oligodendrogliomas, high-grade astrocytomas, and medulloblastomas. Cancer Res. 2004;64(15):5048–50.
112. Samuels Y, Wang Z, Bardelli A, Silliman N, Ptak J, Szabo S, Yan H, Gazdar A, Powell SM, Riggins GJ, Willson JKV, Markowitz S, Kinzler KW, Vogelstein B, Velculescu VE. High frequency of mutations of the PIK3CA gene in human cancers. Science. 2004;304(5670):554.
113. Hurtt MR, Moossy J, Donovan-Peluso M, Locker J. Amplification of epidermal growth factor receptor gene in gliomas: histopathology and prognosis. J Neuropathol Exp Neurol. 1992;51(1):84–90.
114. Knobbe CB, Merlo A, Reifenberger G. Pten signaling in gliomas. Neuro Oncol. 2002;4(3): 196–211.
115. Wong AJ, Ruppert JM, Bigner SH, Grzeschik CH, Humphrey PA, Bigner DS, Vogelstein B. Structural alterations of the epidermal growth factor receptor gene in human gliomas. Proc Natl Acad Sci U S A. 1992;89(7):2965–9.
116. Eyler CE, Foo W-C, LaFiura KM, McLendon RE, Hjelmeland AB, Rich JN. Brain cancer stem cells display preferential sensitivity to Akt inhibition. Stem Cells (Dayton Ohio). 2008;26(12):3027–36.
117. Gallia GL, Tyler BM, Hann CL, Siu I-M, Giranda VL, Vescovi AL, Brem H, Riggins GJ. Inhibition of Akt inhibits growth of glioblastoma and glioblastoma stem-like cells. Mol Cancer Ther. 2009;8(2):386–93.
118. Li H-F, Kim J-S, Waldman T. Radiation-induced Akt activation modulates radioresistance in human glioblastoma cells. Radiat Oncol Lond Engl. 2009;4:43.

119. Park C-M, Park M-J, Kwak H-J, Lee H-C, Kim M-S, Lee S-H, Park I-C, Rhee CH, Hong S-I. Ionizing radiation enhances matrix metalloproteinase-2 secretion and invasion of glioma cells through Src/epidermal growth factor receptor-mediated p38/Akt and phosphatidylinositol 3-kinase/Akt signaling pathways. Cancer Res. 2006;66(17):8511–9.
120. Zhai GG, Malhotra R, Delaney M, Latham D, Nestler U, Zhang M, Mukherjee N, Song Q, Robe P, Chakravarti A. Radiation enhances the invasive potential of primary glioblastoma cells via activation of the Rho signaling pathway. J Neurooncol. 2006;76(3):227–37.
121. Nakamura JL, Karlsson A, Arvold ND, Gottschalk AR, Pieper RO, Stokoe D, Haas-Kogan DA. PKB/Akt mediates radiosensitization by the signaling inhibitor LY294002 in human malignant gliomas. J Neurooncol. 2005;71(3):215–22.
122. Kao GD, Jiang Z, Fernandes AM, Gupta AK, Maity A. Inhibition of phosphatidylinositol-3-OH kinase/Akt signaling impairs DNA repair in glioblastoma cells following ionizing radiation. J Biol Chem. 2007;282(29):21206–12.
123. Kuger S, Graus D, Brendtke R, Günther N, Katzer A, Lutyj P, Polat B, Chatterjee M, Sukhorukov VL, Flentje M, Djuzenova CS. Radiosensitization of Glioblastoma Cell Lines by the Dual PI3K and mTOR Inhibitor NVP-BEZ235 Depends on Drug-Irradiation Schedule. Transl Oncol. 2013;6(2):169–79.
124. Anandharaj A, Cinghu S, Park W-Y. Rapamycin-mediated mTOR inhibition attenuates survivin and sensitizes glioblastoma cells to radiation therapy. Acta Biochim Biophys Sin. 2011;43(4):292–300.
125. Burris 3rd HA. Overcoming acquired resistance to anticancer therapy: focus on the PI3K/AKT/mTOR pathway. Cancer Chemother Pharmacol. 2013;71(4):829–42.
126. Janku F, McConkey DJ, Hong DS, Kurzrock R. Autophagy as a target for anticancer therapy. Nat Rev Clin Oncol. 2011;8(9):528–39.
127. Lomonaco SL, Finniss S, Xiang C, Decarvalho A, Umansky F, Kalkanis SN, Mikkelsen T, Brodie C. The induction of autophagy by gamma-radiation contributes to the radioresistance of glioma stem cells. Int J Cancer. 2009;125(3):717–22.
128. Wullschleger S, Loewith R, Hall MN. TOR signaling in growth and metabolism. Cell. 2006;124(3):471–84.
129. Wang W, Long L, Yang N, Zhang Q, Ji W, Zhao J, Qin Z, Wang Z, Chen G, Liang Z. NVP-BEZ235, a novel dual PI3K/mTOR inhibitor, enhances the radiosensitivity of human glioma stem cells in vitro. Acta Pharmacol Sin. 2013;34(5):681–90.
130. Zhuang W, Li B, Long L, Chen L, Huang Q, Liang Z. Induction of autophagy promotes differentiation of glioma-initiating cells and their radiosensitivity. Int J Cancer. 2011; 129(11):2720–31.
131. Lomonaco SL, Finniss S, Xiang C, Lee HK, Jiang W, Lemke N, Rempel SA, Mikkelsen T, Brodie C. Cilengitide induces autophagy-mediated cell death in glioma cells. Neuro Oncol. 2011;13(8):857–65.
132. Jamal M, Rath BH, Tsang PS, Camphausen K, Tofilon PJ. The brain microenvironment preferentially enhances the radioresistance of CD133(+) glioblastoma stem-like cells. Neoplasia. 2012;14(2):150–8.
133. Cooper LAD, Gutman DA, Chisolm C, Appin C, Kong J, Rong Y, Kurc T, Van Meir EG, Saltz JH, Moreno CS, Brat DJ. The tumor microenvironment strongly impacts master transcriptional regulators and gene expression class of glioblastoma. Am J Pathol. 2012;180(5):2108–19.
134. Barcellos-Hoff MH, Newcomb EW, Zagzag D, Narayana A. Therapeutic targets in malignant glioblastoma microenvironment. Semin Radiat Oncol. 2009;19(3):163–70.
135. Lathia JD, Heddleston JM, Venere M, Rich JN. Deadly teamwork: neural cancer stem cells and the tumor microenvironment. Cell Stem Cell. 2011;8(5):482–5.
136. Heddleston JM, Hitomi M, Venere M, Flavahan WA, Yang K, Kim Y, Minhas S, Rich JN, Hjelmeland AB. Glioma stem cell maintenance: the role of the microenvironment. Curr Pharm Des. 2011;17(23):2386–401.
137. Zeman EM. Biologic basis of radiation oncology, in Clinical radiation oncology, Gunderson LL and Tepper JE, Eds., 3rd ed. Philadelphia, PA Elsevier Saunders; 2012, Chapter 1, pp. 3–42.

138. Denysenko T, Gennero L, Roos MA, Melcarne A, Juenemann C, Faccani G, Morra I, Cavallo G, Reguzzi S, Pescarmona G, Ponzetto A. Glioblastoma cancer stem cells: heterogeneity, microenvironment and related therapeutic strategies. Cell Biochem Funct. 2010;28(5): 343–51.
139. Calabrese C, Poppleton H, Kocak M, Hogg TL, Fuller C, Hamner B, Oh EY, Gaber MW, Finklestein D, Allen M, Frank A, Bayazitov IT, Zakharenko SS, Gajjar A, Davidoff A, Gilbertson RJ. A perivascular niche for brain tumor stem cells. Cancer Cell. 2007;11(1): 69–82.
140. Galli R, Binda E, Orfanelli U, Cipelletti B, Gritti A, De Vitis S, Fiocco R, Foroni C, Dimeco F, Vescovi A. Isolation and characterization of tumorigenic, stem-like neural precursors from human glioblastoma. Cancer Res. 2004;64(19):7011–21.
141. Borovski T, Beke P, van Tellingen O, Rodermond HM, Verhoeff JJ, Lascano V, Daalhuisen JB, Medema JP, Sprick MR. Therapy-resistant tumor microvascular endothelial cells contribute to treatment failure in glioblastoma multiforme. Oncogene. 2013;32(12):1539–48.
142. Zhu TS, Costello MA, Talsma CE, Flack CG, Crowley JG, Hamm LL, He X, Hervey-Jumper SL, Heth JA, Muraszko KM, DiMeco F, Vescovi AL, Fan X. Endothelial cells create a stem cell niche in glioblastoma by providing NOTCH ligands that nurture self-renewal of cancer stem-like cells. Cancer Res. 2011;71(18):6061–72.
143. Gürsel DB, Berry N, Boockvar JA. The contribution of Notch signaling to glioblastoma via activation of cancer stem cell self-renewal: the role of the endothelial network. Neurosurgery. 2012;70(2):N19–21.
144. Charles N, Ozawa T, Squatrito M, Bleau A-M, Brennan CW, Hambardzumyan D, Holland EC. Perivascular nitric oxide activates notch signaling and promotes stem-like character in PDGF-induced glioma cells. Cell Stem Cell. 2010;6(2):141–52.
145. Cheng L, Huang Z, Zhou W, Wu Q, Donnola S, Liu JK, Fang X, Sloan AE, Mao Y, Lathia JD, Min W, McLendon RE, Rich JN, Bao S. Glioblastoma stem cells generate vascular pericytes to support vessel function and tumor growth. Cell. 2013;153(1):139–52.
146. Bao S, Wu Q, Sathornsumetee S, Hao Y, Li Z, Hjelmeland AB, Shi Q, McLendon RE, Bigner DD, Rich JN. Stem cell-like glioma cells promote tumor angiogenesis through vascular endothelial growth factor. Cancer Res. 2006;66(16):7843–8.
147. Oka N, Soeda A, Inagaki A, Onodera M, Maruyama H, Hara A, Kunisada T, Mori H, Iwama T. VEGF promotes tumorigenesis and angiogenesis of human glioblastoma stem cells. Biochem Biophys Res Commun. 2007;360(3):553–9.
148. Bergers G, Benjamin LE. Tumorigenesis and the angiogenic switch. Nat Rev Cancer. 2003;3(6):401–10.
149. Rahmathulla G, Hovey EJ, Hashemi-Sadraei N, Ahluwalia MS. Bevacizumab in high-grade gliomas: a review of its uses, toxicity assessment, and future treatment challenges. OncoTargets Ther. 2013;6:371–89.
150. Gilbert MR, Dignam JJ, Armstrong TS, Wefel JS, Blumenthal DT, Vogelbaum MA, Colman H, Chakravarti A, Pugh S, Won M, Jeraj R, Brown PD, Jaeckle KA, Schiff D, Stieber VW, Brachman DG, Werner-Wasik M, Tremont-Lukats IW, Sulman EP, Aldape KD, Curran Jr WJ, Mehta MP. A randomized trial of bevacizumab for newly diagnosed glioblastoma. N Engl J Med. 2014;370(8):699–708.
151. Zagzag D, Lukyanov Y, Lan L, Ali MA, Esencay M, Mendez O, Yee H, Voura EB, Newcomb EW. Hypoxia-inducible factor 1 and VEGF upregulate CXCR4 in glioblastoma: implications for angiogenesis and glioma cell invasion. Lab Invest. 2006;86(12):1221–32.
152. Zagzag D, Esencay M, Mendez O, Yee H, Smirnova I, Huang Y, Chiriboga L, Lukyanov E, Liu M, Newcomb EW. Hypoxia- and vascular endothelial growth factor-induced stromal cell-derived factor-1alpha/CXCR4 expression in glioblastomas: one plausible explanation of Scherer's structures. Am J Pathol. 2008;173(2):545–60.
153. Ehtesham M, Winston JA, Kabos P, Thompson RC. CXCR4 expression mediates glioma cell invasiveness. Oncogene. 2006;25(19):2801–6.
154. Derynck R, Akhurst RJ, Balmain A. TGF-beta signaling in tumor suppression and cancer progression. Nat Genet. 2001;29(2):117–29.

155. Jachimczak P, Hessdörfer B, Fabel-Schulte K, Wismeth C, Brysch W, Schlingensiepen KH, Bauer A, Blesch A, Bogdahn U. Transforming growth factor-beta-mediated autocrine growth regulation of gliomas as detected with phosphorothioate antisense oligonucleotides. Int J Cancer. 1996;65(3):332–7.
156. Letterio JJ, Roberts AB. Regulation of immune responses by TGF-beta. Annu Rev Immunol. 1998;16:137–61.
157. Peñuelas S, Anido J, Prieto-Sánchez RM, Folch G, Barba I, Cuartas I, García-Dorado D, Poca MA, Sahuquillo J, Baselga J, Seoane J. TGF-beta increases glioma-initiating cell self-renewal through the induction of LIF in human glioblastoma. Cancer Cell. 2009;15(4):315–27.
158. Ikushima H, Todo T, Ino Y, Takahashi M, Miyazawa K, Miyazono K. Autocrine TGF-beta signaling maintains tumorigenicity of glioma-initiating cells through Sry-related HMG-box factors. Cell Stem Cell. 2009;5(5):504–14.
159. Anido J, Sáez-Borderías A, Gonzàlez-Juncà A, Rodón L, Folch G, Carmona MA, Prieto-Sánchez RM, Barba I, Martínez-Sáez E, Prudkin L, Cuartas I, Raventós C, Martínez-Ricarte F, Poca MA, García-Dorado D, Lahn MM, Yingling JM, Rodón J, Sahuquillo J, Baselga J, Seoane J. TGF-β receptor inhibitors target the CD44(high)/Id1(high) glioma-initiating cell population in human glioblastoma. Cancer Cell. 2010;18(6):655–68.
160. Hardee ME, Marciscano AE, Medina-Ramirez CM, Zagzag D, Narayana A, Lonning SM, Barcellos-Hoff MH. Resistance of glioblastoma-initiating cells to radiation mediated by the tumor microenvironment can be abolished by inhibiting transforming growth factor-β. Cancer Res. 2012;72(16):4119–29.
161. Carmeliet P, Jain RK. Angiogenesis in cancer and other diseases. Nature. 2000;407(6801): 249–57.
162. Dewhirst MW, Cao Y, Moeller B. Cycling hypoxia and free radicals regulate angiogenesis and radiotherapy response. Nat Rev Cancer. 2008;8(6):425–37.
163. Evans SM, Jenkins KW, Jenkins WT, Dilling T, Judy KD, Schrlau A, Judkins A, Hahn SM, Koch CJ. Imaging and analytical methods as applied to the evaluation of vasculature and hypoxia in human brain tumors. Radiat Res. 2008;170(6):677–90.
164. Evans SM, Judy KD, Dunphy I, Jenkins WT, Nelson PT, Collins R, Wileyto EP, Jenkins K, Hahn SM, Stevens CW, Judkins AR, Phillips P, Geoerger B, Koch CJ. Comparative measurements of hypoxia in human brain tumors using needle electrodes and EF5 binding. Cancer Res. 2004;64(5):1886–92.
165. Evans SM, Jenkins KW, Chen HI, Jenkins WT, Judy KD, Hwang W-T, Lustig RA, Judkins AR, Grady MS, Hahn SM, Koch CJ. The relationship among hypoxia, proliferation, and outcome in patients with de novo glioblastoma: a pilot study. Transl Oncol. 2010;3(3):160–9.
166. Sathornsumetee S, Cao Y, Marcello JE, Herndon 2nd JE, McLendon RE, Desjardins A, Friedman HS, Dewhirst MW, Vredenburgh JJ, Rich JN. Tumor angiogenic and hypoxic profiles predict radiographic response and survival in malignant astrocytoma patients treated with bevacizumab and irinotecan. J Clin Oncol. 2008;26(2):271–8.
167. Platet N, Liu SY, Atifi ME, Oliver L, Vallette FM, Berger F, Wion D. Influence of oxygen tension on CD133 phenotype in human glioma cell cultures. Cancer Lett. 2007;258(2):286–90.
168. Pistollato F, Abbadi S, Rampazzo E, Persano L, Della Puppa A, Frasson C, Sarto E, Scienza R, D'avella D, Basso G. Intratumoral hypoxic gradient drives stem cells distribution and MGMT expression in glioblastoma. Stem Cells (Dayton Ohio). 2010;28(5):851–62.
169. Persano L, Rampazzo E, Basso G, Viola G. Glioblastoma cancer stem cells: role of the microenvironment and therapeutic targeting. Biochem Pharmacol. 2013;85(5):612–22.
170. McCord AM, Jamal M, Shankavaram UT, Shankavarum UT, Lang FF, Camphausen K, Tofilon PJ. Physiologic oxygen concentration enhances the stem-like properties of CD133+ human glioblastoma cells in vitro. Mol Cancer Res. 2009;7(4):489–97.
171. Majmundar AJ, Wong WJ, Simon MC. Hypoxia-inducible factors and the response to hypoxic stress. Mol Cell. 2010;40(2):294–309.

172. Covello KL, Kehler J, Yu H, Gordan JD, Arsham AM, Hu C-J, Labosky PA, Simon MC, Keith B. HIF-2alpha regulates Oct-4: effects of hypoxia on stem cell function, embryonic development, and tumor growth. Genes Dev. 2006;20(5):557–70.
173. Li Z, Bao S, Wu Q, Wang H, Eyler C, Sathornsumetee S, Shi Q, Cao Y, Lathia J, McLendon RE, Hjelmeland AB, Rich JN. Hypoxia-inducible factors regulate tumorigenic capacity of glioma stem cells. Cancer Cell. 2009;15(6):501–13.
174. Zhao T, Zhang C, Liu Z, Wu L, Huang X, Wu H, Xiong L, Wang X, Wang X, Zhu L, Fan M. Hypoxia-driven proliferation of embryonic neural stem/progenitor cells—role of hypoxia-inducible transcription factor-1alpha. FEBS J. 2008;275(8):1824–34.
175. Simon M-P, Tournaire R, Pouyssegur J. The angiopoietin-2 gene of endothelial cells is up-regulated in hypoxia by a HIF binding site located in its first intron and by the central factors GATA-2 and Ets-1. J Cell Physiol. 2008;217(3):809–18.
176. Yamakawa M, Liu LX, Belanger AJ, Date T, Kuriyama T, Goldberg MA, Cheng SH, Gregory RJ, Jiang C. Expression of angiopoietins in renal epithelial and clear cell carcinoma cells: regulation by hypoxia and participation in angiogenesis. Am J Physiol Renal Physiol. 2004;287(4):F649–57.
177. Bar EE, Lin A, Mahairaki V, Matsui W, Eberhart CG. Hypoxia increases the expression of stem-cell markers and promotes clonogenicity in glioblastoma neurospheres. Am J Pathol. 2010;177(3):1491–502.
178. Zimmer M, Ebert BL, Neil C, Brenner K, Papaioannou I, Melas A, Tolliday N, Lamb J, Pantopoulos K, Golub T, Iliopoulos O. Small-molecule inhibitors of HIF-2a translation link its 5′UTR iron-responsive element to oxygen sensing. Mol Cell. 2008;32(6):838–48.
179. Amundson SA, Bittner M, Chen Y, Trent J, Meltzer P, Fornace Jr AJ. Fluorescent cDNA microarray hybridization reveals complexity and heterogeneity of cellular genotoxic stress responses. Oncogene. 1999;18(24):3666–72.
180. Burns TF, El-Deiry WS. Microarray analysis of p53 target gene expression patterns in the spleen and thymus in response to ionizing radiation. Cancer Biol Ther. 2003;2(4):431–43.
181. Rødningen OK, Overgaard J, Alsner J, Hastie T, Børresen-Dale A-L. Microarray analysis of the transcriptional response to single or multiple doses of ionizing radiation in human subcutaneous fibroblasts. Radiother Oncol. 2005;77(3):231–40.
182. Glantz M, Kesari S, Recht L, Fleischhack G, Van Horn A. Understanding the origins of gliomas and developing novel therapies: cerebrospinal fluid and subventricular zone interplay. Semin Oncol. 2009;36(4 Suppl 2):S17–24.
183. Tada E, Yang C, Gobbel GT, Lamborn KR, Fike JR. Long-term impairment of subependymal repopulation following damage by ionizing irradiation. Exp Neurol. 1999;160(1):66–77.
184. Panagiotakos G, Alshamy G, Chan B, Abrams R, Greenberg E, Saxena A, Bradbury M, Edgar M, Gutin P, Tabar V. Long-term impact of radiation on the stem cell and oligodendrocyte precursors in the brain. PloS One. 2007;2(7):e588.
185. Monje ML, Mizumatsu S, Fike JR, Palmer TD. Irradiation induces neural precursor-cell dysfunction. Nat Med. 2002;8(9):955–62.
186. Redmond KJ, Mahone EM, Terezakis S, Ishaq O, Ford E, McNutt T, Kleinberg L, Cohen KJ, Wharam M, Horska A. Association between radiation dose to neuronal progenitor cell niches and temporal lobes and performance on neuropsychological testing in children: a prospective study. Neuro Oncol. 2013;15(3):360–9.
187. Llaguno SA, Chen J, Kwon C-H, Parada LF. Neural and cancer stem cells in tumor suppressor mouse models of malignant astrocytoma. Cold Spring Harb Symp Quant Biol. 2008;73:421–6.
188. Holland EC, Celestino J, Dai C, Schaefer L, Sawaya RE, Fuller GN. Combined activation of Ras and Akt in neural progenitors induces glioblastoma formation in mice. Nat Genet. 2000;25(1):55–7.
189. Uhrbom L, Dai C, Celestino JC, Rosenblum MK, Fuller GN, Holland EC. Ink4a-Arf loss cooperates with KRas activation in astrocytes and neural progenitors to generate glioblastomas of various morphologies depending on activated Akt. Cancer Res. 2002;62(19):5551–8.

190. Bachoo RM, Maher EA, Ligon KL, Sharpless NE, Chan SS, You MJ, Tang Y, DeFrances J, Stover E, Weissleder R, Rowitch DH, Louis DN, DePinho RA. Epidermal growth factor receptor and Ink4a/Arf: convergent mechanisms governing terminal differentiation and transformation along the neural stem cell to astrocyte axis. Cancer Cell. 2002;1(3):269–77.
191. Zhu Y, Guignard F, Zhao D, Liu L, Burns DK, Mason RP, Messing A, Parada LF. Early inactivation of p53 tumor suppressor gene cooperating with NF1 loss induces malignant astrocytoma. Cancer Cell. 2005;8(2):119–30.
192. Jackson EL, Garcia-Verdugo JM, Gil-Perotin S, Roy M, Quinones-Hinojosa A, VandenBerg S, Alvarez-Buylla A. PDGFR alpha-positive B cells are neural stem cells in the adult SVZ that form glioma-like growths in response to increased PDGF signaling. Neuron. 2006;51(2):187–99.
193. Glass R, Synowitz M, Kronenberg G, Walzlein J-H, Markovic DS, Wang L-P, Gast D, Kiwit J, Kempermann G, Kettenmann H. Glioblastoma-induced attraction of endogenous neural precursor cells is associated with improved survival. J Neurosci. 2005;25(10):2637–46.
194. Duntsch C, Zhou Q, Weimar JD, Frankel B, Robertson JH, Pourmotabbed T. Up-regulation of neuropoiesis generating glial progenitors that infiltrate rat intracranial glioma. J Neurooncol. 2005;71(3):245–55.
195. Gibbs IC, Haas-Kogan D, Terezakis S, Kavanagh BD. The subventricular zone neural progenitor cell hypothesis in glioblastoma: epiphany, Trojan Horse, or Cheshire fact? Int J Radiat Oncol Biol Phys. 2013;86(4):606–8.
196. Jafri NF, Clarke JL, Weinberg V, Barani IJ, Cha S. Relationship of glioblastoma multiforme to the subventricular zone is associated with survival. Neuro Oncol. 2013;15(1):91–6.
197. Chaichana KL, McGirt MJ, Frazier J, Attenello F, Guerrero-Cazares H, Quinones-Hinojosa A. Relationship of glioblastoma multiforme to the lateral ventricles predicts survival following tumor resection. J Neurooncol. 2008;89(2):219–24.
198. Chen L, Guerrero-Cazares H, Ye X, Ford E, McNutt T, Kleinberg L, Lim M, Chaichana K, Quinones-Hinojosa A, Redmond K. Increased subventricular zone radiation dose correlates with survival in glioblastoma patients after gross total resection. Int J Radiat Oncol Biol Phys. 2013;86(4):616–22.
199. Lee P, Eppinga W, Lagerwaard F, Cloughesy T, Slotman B, Nghiemphu PL, Wang P-C, Kupelian P, Agazaryan N, Demarco J, Selch MT, Steinberg M, Kang JJ. Evaluation of high ipsilateral subventricular zone radiation therapy dose in glioblastoma: a pooled analysis. Int J Radiat Oncol Biol Phys. 2013;86(4):609–15.

Chemoresistance and Chemotherapy Targeting Stem-Like Cells in Malignant Glioma

Mia Dahl Sørensen, Sigurd Fosmark, Sofie Hellwege, Dagmar Beier, Bjarne Winther Kristensen, and Christoph Patrick Beier

Abstract Glioblastoma remains a tumor with a dismal prognosis because of failure of current treatment. Glioblastoma cells with stem cell (GSC) properties survive chemotherapy and give rise to tumor recurrences that invariably result in the death of the patients. Here we summarize the current knowledge on chemoresistance of malignant glioma with a strong focus on GSC. Chemoresistant GSC are the most likely cause of tumor recurrence, but it remains controversial if GSC and under which conditions GSC are more chemoresistant than non-GSC within the tumor. Regardless of this uncertainty, the chemoresistance varies and it is mainly mediated by intrinsic factors. O6-methyl-guanidine methyltransferase (MGMT) remains the most potent mediator of chemoresistance, but disturbed mismatch repair system and multidrug resistance proteins contribute substantially. However, the intrinsic resistance by MGMT expression is regulated by extrinsic factors like hypoxia increasing MGMT expression and thereby resistance to alkylating chemotherapy. The search of new biomarkers helping to predict the tumor response to chemotherapy is ongoing and will complement the already known markers like MGMT.

Keywords Glioblastoma stem cell • Cancer stem cell • Temozolomide • MGMT • MMR • MDR • Glioblastoma

M.D. Sørensen • S. Fosmark • S. Hellwege • B.W. Kristensen
Department of Pathology, Odense University Hospital, Odense C, Denmark

Institute of Clinical Research, University of Southern Denmark, Odense M, Denmark

D. Beier • C.P. Beier, M.D. (✉)
Department of Neurology, University of Southern Denmark, Sdr. Boulevard 19, 5000 Odense C, Denmark

Institute of Clinical Research, University of Southern Denmark, Odense M, Denmark
e-mail: cbeier@health.sdu.dk

M. Ehtesham (ed.), *Stem Cell Biology in Neoplasms of the Central Nervous System*, Advances in Experimental Medicine and Biology 853,
DOI 10.1007/978-3-319-16537-0_7

Introduction

Glioblastoma multiforme (GBM) is the most common primary brain tumor and the prognosis of affected patients remains dismal in spite of recent improvements of chemo- and supportive therapy. The average survival of patients receiving optimized multimodal treatment including resection, radiotherapy, and concomitant and adjuvant chemotherapy with temozolomide as well as supportive treatment is about 15 months [1, 2]. The average survival of all patients is even lower [3]. GBM is unique among the most aggressive tumors in humans, because they hardly metastasize. However, they show all signs of malignancies including invasive growth and marked resistance to therapy [4]. The major pillar of today's treatment remains radiotherapy constituting the most effective part of all current therapeutic regimes. However, it was not possible to substantially improve the therapeutic effect of radiotherapy despite multiple approaches aiming at optimization of treatment protocols. Current research therefore mainly focuses on the reduction of side effects [5]. Thus, new and more effective chemotherapeutic strategies are needed. GBM displays a substantial resistance to a multitude of chemotherapeutic drugs and the alkylating drugs nitrosourea-derivates (ACNU, BCNU, and CCNU) and temozolomide are the only substances with a proven efficacy against GBM.

Gliomas are subclassified into astrocytic, oligodendroglial, oligoastrocytic ("mixed gliomas"), and ependymal tumors according to their histological similarities to different types of glial cells [4]. Moreover, gliomas are subdivided according to their malignancy in WHO I–IV tumors with GBM corresponding to IV tumors. GBM is clinically further divided into primary and secondary GBM dependent on its origin with secondary GBM developing out of II–III gliomas [6]. On a molecular level, GBM comprises at least four different subgroups that do not show any clinical significant differences with response to standard therapy based on surgery, radiotherapy, and chemotherapy with TMZ [7–9].

GBM is driven by tumor cells with stem cell properties (GBM stem cells, GSC) that show high similarities with neural stem cells [10]. They possess all properties of somatic stem cells including infinite growth, asymmetric cell division, and multipotency (differentiation into tumor cells showing properties of all neural lineages), and are able to reconstitute the original tumor after transplantation into nude mice [11–13].

Although GSC and NSC share common properties, e.g., stem cell properties, it is still unclear if NSC represent the cell of origin of GSC [14]. Alternative hypotheses include among others GSC evolving from more differentiated cells via acquisition of stem cell properties and dedifferentiation [15].

The Cancer Stem Cell Model and Chemoresistance

The initial concept of cancer stem cells (CSC) postulated that each cancer is organized hierarchically with CSC at the top of the hierarchy [16] (Fig. 1). Although this model was already proposed in the 1970s [17], it became very popular in recent

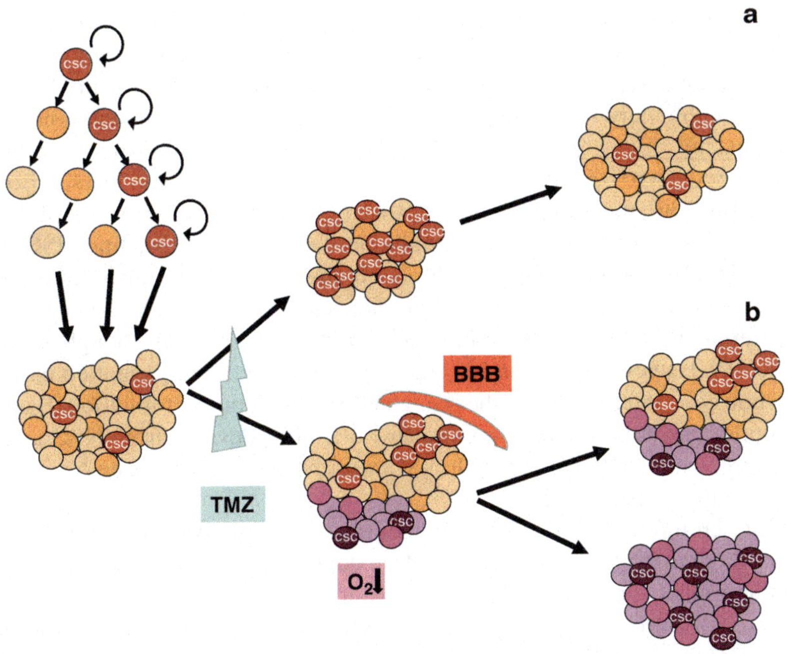

Fig. 1 Cancer stem cell models of chemoresistance. An overview over the classical model of chemoresistance (**a**) and a modified model (**b**) is given (*BBB* blood-brain barrier)

years due to the availability of suitable markers allowing the prospective isolation of the CSC. It turned out that the model did not fit to all tumors because in some solid tumors—e.g., melanoma—the vast majority of tumor cells may have stem cell properties. The model was therefore recently revised and a modified model tried to include recent data [18].

GBM cells with stem cell properties was first isolated in 2002 by Ignatova et al. [12] and these findings could be reproduced by many groups in almost all GBM [11, 13, 19–27]. The cell of origin of GBM (like other astrocytic tumors) and thereby of GSC remains unknown [14]. The name of the respective tumors derives from histological similarities with the respective glial cells. Although astrocytes (and even neurons) can be transformed into GBM-like tumors in different mouse models [28–30], tumors develop from NSC in most animal models with spontaneous GBM formation [30–33] and NSC were therefore considered to be the putative cells of origin of GBM GSC.

NSC are a small subpopulation of cells in the neurogenic regions in the adult brain around the ventricles (subventricular zone) and in the hippocampus. They reside in neurogenic niches that have been extensively characterized recently. Briefly, NSC give rise to rapidly proliferating progenitor cells that further develop a.o. to dedicate cells like neuroblast and glial progenitor cells. These neuroblasts migrate in rodents – and most likely also in humans – to their target structures in the bulbus olfactorius and the hippocampus (overview in: [34, 35]). One important detail of neurogenesis is the strict hierarchy with cells developing (under physiological conditions) always unidirectional from undifferentiated NSC—with the ability to proliferate infinitely—towards more differentiated cells with limited or no proliferation capacity.

GSC turned out to be very similar to NSC. Because NSC show a marked tropism towards GBM [36], the initial publications were asked to prove by the use of genetic markers that the cells investigated actually are GSC and not invading NSC. Both NSC and GSC require similar growth conditions and spontaneously differentiate into cells expressing markers of glial, neuronal, and oligodendroglial lineages. They even differentiate partially into electrically active neurons with though impaired function [10]. However, GSC formed GBM-like tumors after implantation into nude mice, while NSC integrated into the brain tissue without tumor formation. Based on these similarities, it was tempting to assume that GSC develop from NSC and that GSC share common properties with NSC with respect to resistance to cytotoxic drugs (e.g., chemotherapy).

In NSC cultures, the use of chemotherapeutic drugs like ARA-C selectively kills rapidly proliferating cells and therefore depletes all progenitors with the stem cell compartment remaining and being able to recapitulate the cell culture [37]. This approach therefore allows isolating the bona fide stem cells by their relative chemoresistance to ARA-C. Another approach is also based on the low proliferation rate of NSC. NSC are labelled by a modified uridine (BrdU) being incorporated into the DNA. While proliferating cells distribute labelled BrdU to their daughter cells, NSC maintain BrdU and are therefore named as "labelling retaining cells." BrdU-retaining cells are often considered to be equivalent to bona fide stem cells.

The relative resistance of NSC to ARA-C therefore gave rise to the hypothesis that GSC are also relatively chemoresistant and are thus the likely cause of chemoresistance in GBM. This concept was further supported by experiments using the fluorochrome Hoechst 33342. NSC removed Hoechst 33342 from the cell soma ("side population") due to the expression of transporter proteins (MDR1/p-glycoprotein, ABCG2) that were also able to remove chemotherapeutic drugs from the cell soma ("multidrug resistance proteins," see below, [38, 39]).

This leads to the hypothesis of GSC being more resistant to chemotherapy than non-stem cells, and of GSC therefore constituting the major mediator of resistance to chemotherapy and tumor recurrences [16]. The transient effect of the chemotherapy with tumor shrinkage was attributed to the death of rapidly proliferating progenitor cells without stem cell properties that were postulated to be the main targets of (chemo-)therapy. According to this hypothesis, the lack of current chemotherapies to eliminate the bona fide stem cells is the reason for chemoresistance. This model became very popular and remains a pillar stone in the field's perception of GSC and chemoresistance (Fig. 1).

Major Problems of the Classical CSC Model of Chemoresistance

However, in recent years, several inconsistencies in this otherwise very intriguing model became apparent raising doubts doubts if this model—at least in this oversimplified version—holds true in GBM.

Cellular Hierarchy

The classical model of chemoresistance due to GSC postulates and requires a unidirectional development from stem cells to progenitor cells to differentiated cells [16]. This implicates that the elimination of the bona fide stem cells, and thereby the top of the hierarchy, will be sufficient to eliminate tumor growth. However, a detailed analysis of the GSC cultures in recent years unveiled that the hierarchical model is clearly an oversimplification in most GBM [40].

Differentiated Cells Can Acquire Stem Cell Properties

Differentiated cells can acquire stem cell properties, e.g., by hypoxia-induced HIF 2α expression [41]. This concept is supported by a recent publication reporting the selective elimination of all cells expressing "nestin" from tumor in a mouse model of spontaneous GBM. Nestin is an established marker with broad expression in stem and progenitor cells but with low specificity. Even though the selective elimination was successful, the elimination did only transiently reduce the growth of the tumor but could not completely stop tumor growth [42]. Because the recurrent tumors derived from nestin-positive GSC, this raises substantial doubts on the presence of a unidirectional hierarchy in GBM. Likewise could Auffinger et al. show that sublethal concentrations of chemotherapy may induce stemness in differentiated cells [43].

Different Types of GSC Coexist Within GBM

Substantial evidence suggests that GBM may comprise different types of GSC. While subgroups of them may be hierarchically organized, they are also able to transform into one another dependent on the local environment [40].

The Heterogeneity of GSC Reflects the Heterogeneity of GBM

The molecular diversity of GBM on a cellular, RNA, DNA, and protein level [9] is reflected by the heterogeneity of GSC. Although all groups used different approaches

[7, 8, 44], all groups succeeded to describe two groups of GSC resembling proneural (PN) and mesenchymal (Mes) GBM. These subtypes may transform into each other as suggested by a recent publication of Bhat et al. [45].

Progenitor Cells

The classical model of chemoresistance requires a hierarchy with slowly proliferating, chemoresistant GSC at the top of the hierarchy and rapidly proliferating, chemo-susceptible progenitor cells as intermediate-stage cells. Although there is no doubt about the existence of GBM cells with stem cell properties, the other parts of the hierarchy are only partially characterized. It is among the most surprising facts in the research on GSC that the postulated progenitor cell compartment, despite representing a crucial part of the initial model, is hardly characterized. The most likely explanation is the lack of valid markers to prospectively isolate these cells. So far, there is only indirect evidence for the existence of progenitor cells from a study investigating a hierarchically organized GSC line. Here, the loss of telomere length and differences in the telomerase expression suggested the presence of $CD133^{-}$/telomerase^{-} progenitors [46].

Clinical Data

The classical CSC model postulates shrinkage of the tumor after therapy by preferential elimination of non-stem cells. However, in the clinical practice, GBM hardly ever shrinks after standard radiochemotherapy [47]. The tumor may remain stable over several months without treatment. Even if therapy would substantially affect the progenitor cell pool, this can hardly explain the transient effect of therapy without relevant tumor shrinkage.

Cancer Stem Cells and Chemoresistance

In spite of the doubts regarding the applicability of the CSC model to GBM, there is consensus that GSC (i.e., stem cell-like GBM cells) explain recurrence after chemotherapy. In the last 10 years, many laboratories have provided data on the effects of chemotherapy on GSC. Almost all publications used TMZ for their experiments and there is only enough data on TMZ and GSC allowing to distillate what is accepted knowledge and which topics are under discussion or unresolved.

One major controversy focused on if all GSC are chemoresistant to all drugs. The experimental data provided in the last 7 years is conflicting and does not allow a simple "yes/no" answers to this question—the reality turned out to be much more complex than previously thought [48].

Due to theoretical considerations, there is broad consensus that tumor recurrences after chemotherapy develop from cells with stem cell properties. Data in support of this concept were provided by mouse models and patient samples.

Using a murine mouse model with spontaneous development of GBM, Chen et al. could prove that cells expressing the stem cell marker nestin are crucial for tumor recurrence after TMZ treatment. In this model, GBM-like tumors developed spontaneously from the neurogenic regions in the brain. NSC residing in these regions are characterized by the expression of the neurofilament nestin. The parallel expression of GFP and thymidine kinase under the nestin promoter allowed the selective identification and elimination of nestin-expressing cells using ganciclovir. After TMZ treatment, both the number of proliferating cells and the number of stem cells were reduced substantially and recurrent tumor developed from stem cell-like tumor cells that started to proliferate again as early as 5 days after TMZ treatment. The treatment of ganciclovir substantially reduced tumor growth but could not completely eradicate the tumor [42].

Similar findings were found in patient samples. Pallini et al. compared the stem cell content of tumor probes before combined radiochemotherapy and of recurrent tumors [49]. They could show that the proportion of tumor cells expressing the stem cell marker CD133 increases up to 20× after combined radiochemotherapy suggesting an enrichment of stem cell-like cells. This clearly suggests an enrichment of stem cell-like cells in recurrent brain tumors after therapy and supports the concept of stem cells as a mediator of tumor recurrence [50].

There is a broad consensus that CSC are more resistant to radiotherapy than cells without stem cell properties. Different groups could reproduce the key experiments and several mechanisms are postulated mediators of this resistance (e.g., [21, 51]). However, it is more controversial if stem cells are actually more resistant to chemotherapy with TMZ as compared to cells without stem cell properties within the same tumor [52–54]. The next paragraph describes results of several experiments that have been provided by different laboratories. However, they do not allow a definite answer a.o. because of the lack of perfect stem cell markers for in vitro and in vivo research.

A series of publications consistently reported that GSC (cultured under serum-free culture conditions) displayed an increased resistance to chemotherapy as compared to GSC cultured under culture condition favoring tumor cell differentiation (with serum, [23, 55]). However, the culture conditions also change the growth pattern of the cells (sphere formation when culture using serum-free "stem cell medium"), which *per se* confers resistance to TMZ. So far, it was not technically possible to perfectly control for "growth pattern" as a relevant biasing factor [56]. Conversely, all laboratories reported that pulse treatment with TMZ rapidly and substantially eliminates clonogenic cells from in vitro cultures and may eliminate growth completely without killing the cells in some GSC lines [52, 57–59]. TMZ pulse treatment also reduced the expression of stem cell markers like nestin and CD133 suggesting that stem cells were eliminated [52, 60]. This effect appears robust and reproducible.

However, cells that survived a single-pulse treatment with TMZ give rise to cell culture that closely resembles the initial cultures suggesting that GSC survived [61] and that the elimination of clonogenic cells (and stem cell marker expression) may have been mainly due to elimination of postulated progenitor cells. Due to the lack of nonambiguous and unique stem cell markers it was not possible at the moment to precisely prove if TMZ has a stronger effect on stem cells as compared to non-stem cells, especially the group of postulated progenitor cells. A final interpretation of the described effects is therefore pending.

In contrast, chronic treatment with TMZ enriches for tumor cells with stem cell properties, induces stemness, and thereby increases the tumor aggressiveness [43, 53, 61]. Using a murine model of glioblastoma, Bleau et al. could show that long-term treatment with subtoxic concentration of the TMZ substantially increases tumorigenicity and tumor growth [53]. Auffinger et al. detected an induction of stem cell properties in differentiated glioma cells after TMZ treatment [43]. In line with these results, in vitro treatment with sublethal TMZ concentration for 21 days increased both the proportion of CD133-expressing and clonogenic cells. It remains difficult to interpret data from a recent clinical trial comparing different dosing schemes: the authors did not find a difference between shorter treatment (5 of 28 days) and more chronic treatment (21 of 28 days with a lower daily dose) neither with respect to progression-free survival nor with overall survival [61].

A recent paper now proposed that there is no difference between stem cells and non-stem cells to combined radiochemotherapy, although the technical approach used in this publication bore several limitations [62].

Concept of Intrinsic vs. Extrinsic Resistance of GSC

Irrespective of the intriguing question if GSC are more chemoresistant than more differentiated tumor cells and if the cellular hierarchy remains intact in GBM, there is no doubt that therapies aiming at the prevention of tumor recurrence have to successfully eliminate all GSC (and likely all other cells with the ability to acquire CSC properties). It is therefore of outstanding importance to understand the mechanisms providing chemoresistance to GSC.

GSC may resist chemotherapy due to endogenous resistance mechanism ("intrinsic chemoresistance") like the overexpression of MGMT. The chemoresistance may also be dependent on extrinsic factors like metabolic conditions in the microenvironment (e.g., hypoxia) or be caused by aberrant activation of signaling pathways (e.g., EGF-PI3 kinase-AKT). This "extrinsic chemoresistance" may effectuate its effects via activation of proteins responsible for detoxification of TMZ. Alternative extrinsic factors may modulate proliferation or state (e.g., induction of dormancy) of cells. Extrinsic chemoresistance is reversible and therefore potentially targetable [48, 63].

Mechanisms of Chemoresistance

Commonly Used Chemotherapeutic Agents in GBM

There are many different chemotherapeutic drugs that have been investigated in vain in phase II and III studies in recent years [64]. At the moment, there are two substances with proven effectiveness to GBM [1, 2, 65–67]:

BCNU (carmustine) belongs to a group of alkylating substances derived from World War I mustard gas (including nimustine (ACNU) and lomustine (CCNU)—nitrosoureas, CNUs). They alkylate mainly at the N7-position of guanine and the N3-position of adenine. Further CNUs chloroethylate the O6-position of guanine which cross-links DNA due to an N^1-deoxyguanosinyl-N^3-deoxycytidyl resulting in direct strand breaks. These strand breaks induce apoptosis in cells with functional p53, while cells with defective p53 system undergo necrotic cell death [68–70].

TMZ belongs to a group of alkylating agents that readily cross the blood-brain barrier (BBB), and may reach therapeutic concentrations within the brain parenchyma. TMZ requires activation by nonenzymatic conversion at physiologic pH to its reactive compound 5-3-(methyl)-1-(triazen-1-yl) imidazole-4-carboxamide (MTIC). MTIC causes DNA damage mainly by methylating the O^6-position of guanine (primary lesion). Methylated guanine then mismatches with thymine in double-stranded DNA (O^6G-T) in the first cell cycle after treatment. This mismatch induces futile cycles of mismatch repair triggered by recurrent GT mismatches resulting in either double-strand breaks or a critical recombinogenic secondary lesion (Fig. 2). An apurinic/athymidinic site formed during faulty mismatch repair is usually the secondary lesion that either blocks replication (i.e., G2/M cell cycle arrest) or results in apoptotic cell death due to tertiary lesions like DNA double-strand breaks, sister chromatid exchange, or other aberrations [70]. Thus, it is not the primary lesions caused by TMZ, but the tertiary lesions formed during faulty mismatch repair, that induce cell cycle arrest or cell death of the affected tumor cells (from [48]).

With respect to GSC, there is only sufficient data available on TMZ with only few papers investigating resistance to BCNU. The following paragraphs therefore focus exclusively on mechanisms mediating resistance to TMZ and CNUs. Due to the paucity of data we will not mention the few reports on GSC and other drugs sporadically used in therapy of GBM like irinotecan [64].

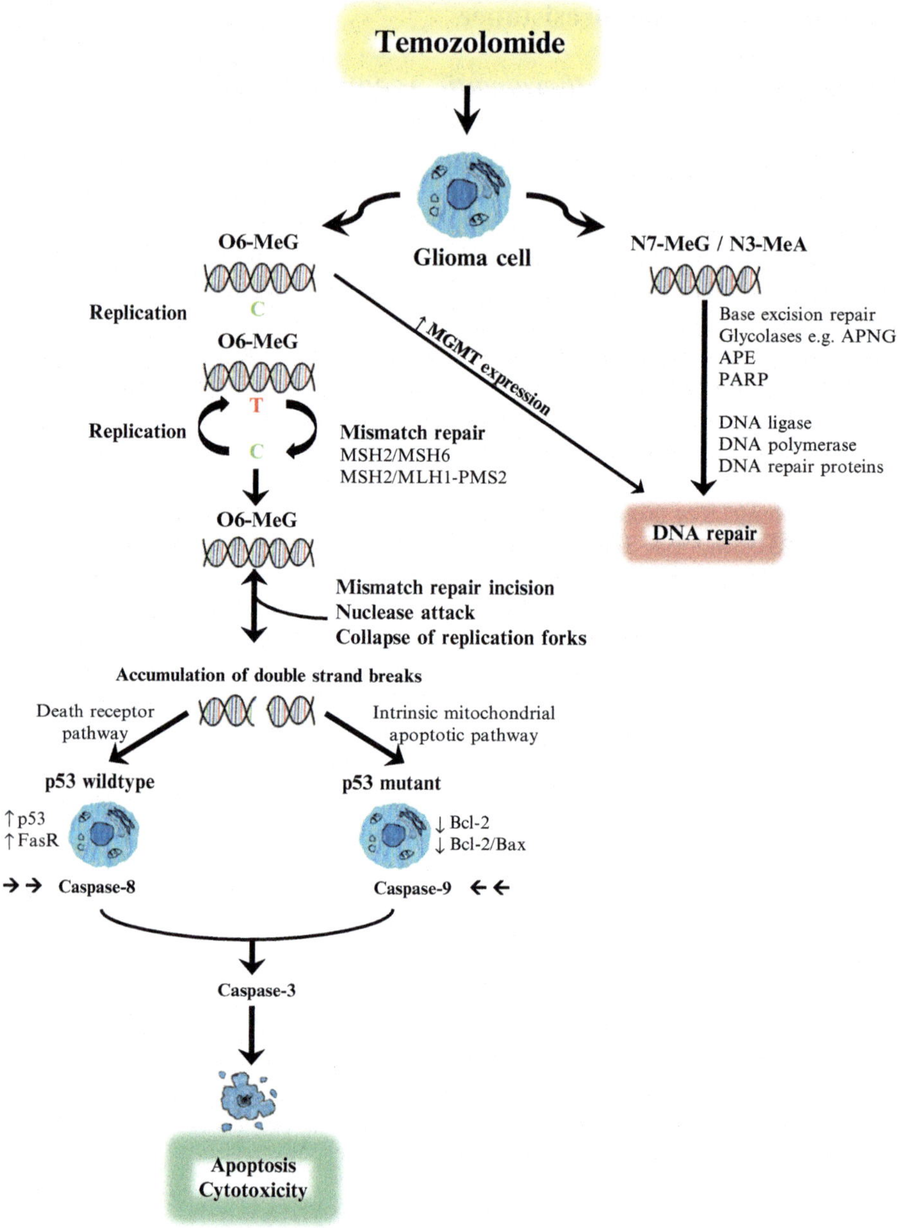

Fig. 2 Mechanism of TMZ toxicity. TMZ primary cytotoxic effect is methylation (O6-MeG) of guanine at the O^6-position, which leads to mismatch with thymine in the double-stranded DNA. This mispairing activates the DNA mismatch repair (MMR) system leading to futile cycles of DNA repair in the end causing double-strand breaks and induction of apoptosis. Apoptosis may be mediated through p53-CD95L/FasL receptor pathway with activation of caspase-8, or triggered by a p53 independent pathway with diminished Bcl-2/Bax ratio and activation of caspase-9. Both pathways result in the activation of the apoptotic cascade and cell death. Figure is based on [70]

Mechanism of Intrinsic Resistance

Mechanisms of Action of MGMT

Added methyl groups due to CNU and TMZ can be removed by the DNA repair protein O^6-methylguanine-DNA-methyl-transferase (MGMT). It removes, in particular, the O6 adducts from guanine by transferring the added methyl groups to a cysteine at its active site and binding it covalently to its sulfur atom. The protein is itself inactivated and further released from the DNA for breakdown. Thus stoichiometrically, MGMT operates in a suicide fashion and has the ability to reverse the effect caused by the mentioned chemotherapeutic agents [71]. MGMT is consequently working to protect the cell from DNA damage, by reversing carcinogenic lesions caused by alkylation. However, since early studies suggested an association between MGMT-negative cells and reduced survival when treated with methylating agents, the protein has been suspected as a culprit in making certain tumors highly resistant to chemotherapy [72].

Located on chromosome 10q26, the MGMT gene has a promoter containing a CpG island of 97 cytosine-guanine repeats [73–77]. Methylations of some of these cytosine bases and the ensuing alteration of chromatin structure have been shown to suppress the expression of MGMT. But there are several sites at which methylation also may occur, all of which have not shown to be linked with the downregulation of the gene [73, 78].

MGMT is ubiquitously expressed in many cells, and seemingly in great excess compared with the frequency at which its targeted lesions occur. Although expression levels vary, liver tissue is typically exhibiting the highest and hematopoietic and brain tissue the lowest expression levels [71, 76, 77, 79]. Surprisingly, NSC express MGMT to a similar amount as GSC [80]. It has been shown that MGMT is expressed in certain cancer cells, often to a greater extent than in the surrounding normal tissue, and especially in gliomas, the protein has shown to act as a clinically relevant biomarker [81].

MGMT is also the best characterized mediator of chemoresistance in GSC. MGMT expression in GSC is also—but not exclusively—regulated by epigenetic modification of the MGMT promoter. The methylation of CpG islands of the promoter region correlates with the response of the tumor with TMZ and—though with some limitations—with the expression of MGMT and the activity of MGMT in tumor cells [82, 83]. Interestingly, the CpG methylation patterns were characteristic for each glioblastoma and did not change when GSC were cultured in vitro. A comparison with the patient samples revealed a patient-specific pattern of methylation; however, the percentage of cells showing this methylation pattern dramatically increased in vitro [84]. This suggest on one hand, that GSC cultures are enriched with cells with methylated MGMT promoter, and, on the other hand, that many non-GSC cells in the GBM do not have a methylated MGMT promoter. This is of high relevance, because—as expected and in line with other publications—the methylation pattern correlated with MGMT expression in GSC cultures in vitro. It increases the resistance of GSC up to 10× [52] and its role in mediating chemoresistance has repeatedly been confirmed [59, 78, 81, 85].

It remains controversial if GSC express more or less MGMT as compared to non-GSC. Due to the abovementioned controversies on the GSC model per se and the methodological problems of all studies so far, a final assessment is not possible yet.

Mismatch Repair Genes

The mismatch repair (MMR) gene pathway is crucial in mediating the cytotoxic effect of O6-methylguanine [86]. Several proteins constitute the MMR machinery (Table 1). Its function is to recognize and correct DNA base mismatch generated in the DNA replication, which has escaped polymerase proofreading. DNA polymerase

Table 1 Overview over known mechanisms of resistance against alkylating chemotherapy in glioblastoma

Group	Members associated with chemoresistance	Description
DNA methyl-transferase	MGMT	DNA methyltransferases are an enzyme family belonging to the group DNA repair proteins.
		MGMT repairs DNA lesions through covalent transfer of the alkyl from the O6-alkylguanine DNA adduct group to the conserved active site, cysteine restoring the guanine to normal.
ABC superfamily	ABCB1 (MDR1, P-gp, CD243)	ATP-binding cassette transporters are ATP-dependent efflux pumps, capable of translocating endogenous compounds, e.g., lipids and xenobiotics including chemotherapeutic agents across extra- and intracellular membranes
	ABCC1 (MRP1)	ABC genes are divided into seven subfamilies A-G (ABC1, MDR/TAP, MRP/CFTR, ALD, OABP, GCN20, White)
	ABCG2 (BCRP/MXR1/ABCP)	ABCB1 is a member of the MDR/TAP subfamily, ABCC1 and ABCC3 of the MRP/CFTR subfamily
	ABCC3 (CTGF)	ABCG2 of the White subfamily
Vaults	LRP/MVP	MVP is a component in the vault complex
		Vaults are the largest ribonucleoprotein particles, ubiquitously expressed transporting substances involved in normal cell function such as hormones, ribosomes, and mRNA
		Possibly involved in multidrug resistance through a nucleo-cytoplasmatic transport, mainly translocating DNA-targeting drugs
BCL2 protein family	Bax, Bak, Bcl-X_S	Bax, Bak, and Bcl-X_S are inducers of apoptosis (pro-apoptotic regulators)
	Bcl-2, Bcl-X_L	Bcl-2 and Bcl-X_L are suppressors of apoptosis (anti-apoptotic regulators)
IAP	XIAP, cIAP1, cIAP2, NAIP	Anti-apoptotic proteins blocking and inhibiting caspases 3, 7, and 9, thereby preventing apoptosis

(continued)

Table 1 (continued)

Group	Members associated with chemoresistance	Description
MMR	MutSα (MSH2 and MSH6)	MMR gene complexes recognize and bind to mismatch lesions
	MutSβ (MSH2 and MSH3)	In humans Mutsα is the most important
	Mut1Lα (MLH1-PMS2)	The MSH2/MSH6 heterodimer recruits the MLH1/PMS2 heterodimer which organizes the interaction between the MMR complex and additional proteins (nucleases, e.g., exonuclease-1, helicase, DNA polymerase ε/δ, and ligases) needed for removal and replacement of mismatched DNA bases
	Mut1Lβ (MLH1-PMS1)	
BER	PARP-1	PARP-1 facilitates efficient DNA repair by modifying nuclear proteins involved in the BER pathway, but is also vital in other cellular processes, e.g., apoptosis
	APE-1	PARP-1 is constitutively expressed, but activated in response to DNA damage including double-strand breaks. Other components necessary in this machinery are lesion-specific glycolases, e.g., APNG/MPG and APE-1, a multifunctional enzyme with both a DNA repair domain and a redox domain
		In the BER pathway APE-1 cleaves the phosphodiester backbone on the 5′ side of the AP (apurinic/apurimidinic site, generated by the glycolases), leaving the 3′-OH and 5′ deoxyribose phosphate termini at the DNA strand break. Also involved in redox-sensitive activation of transcription factors such as p53, NFκβ, AP-1, and HIF-1α
ALDH	ALDH1A1	ALDH is an enzyme family comprising 19 isoforms localized in the cytoplasm, mitochondria, or nucleus-oxidizing aldehydes to carboxylic acids. Aldehydes are generated by various metabolic processes including lipid peroxidation, but also from catabolism of amino acids, carbohydrates, and neurotransmitters
	ALDH3A1	ALDH functions include detoxification and antioxidation
		Regarding cancers ALDH1A1 and ALDH3A1 are considered possible biomarkers

Abbreviations: *ABC* adenosine triphosphate-binding cassette, *ALDH* aldehyde dehydrogenase, *AP* activator protein, *APE* apurinic-apyrimidinic endonuclease, *APNG* alkylpurine-DNA-*N*-glycolase, *Bak* Bcl-2-antagonist/killer, *Bax* Bcl-2-associated X protein, *BCL2* B-cell lymphoma 2, *BER* base excision repair, *CD* cluster of differentiation, *CFTR* cystic fibrosis transmembrane conductance regulator, *HIF* hypoxia inducible factor, *IAP* inhibitor of apoptosis protein, *LRP* lung resistance-related protein, *MDR* multidrug resistance, *MMR* mismatch repair, *MPG* *N*-methylpurine-DNA glycosylase, *MPR* multidrug resistance like protein, *MSH* mutS homolog, *MLH* mutL homolog, *Muts MutL* mutator S L, *MVP* major vault protein, *NFκβ* nuclear factor kappa B, *PARP* poly-(ADP-ribose)polymerase, *TAP* transporter associated with antigen processing, *P-gp* P-glycoprotein, *PMS* postmeiotic segregation increased mispairs O6-methylguanine (induced by TMZ and other alkylating agents) with thymine (instead of cytosine) during DNA replication alerting the MMR system to remove thymine on the daughter strand. The TMZ-induced O6-methylguanine, however, continues to exist within the template strand resulting in successive rounds of mismatch repair. Eventually, the MMR system induces double-strand breaks with replication fork collapse followed by cell cycle arrest and apoptosis [86]. Consequently, dysfunctional MMR leads to cells that are relatively TMZ resistant as they acquire a tolerance to the mismatch of O6-methylguanine with thymine and continue to survive and divide despite DNA damage and at the expense of increased mutagenesis [86, 87]. In fact, MMR-deficient cells have been reported to be up to 100-fold less sensitive to alkylating agents compared to their MMR wild-type counterparts [87]. Somatic hypermutations have been found in several genes involved in MMR in samples derived from GBM patients treated with TMZ (mutator L and mutator S homologs; MLH1, MSH2, MSH6, PMS2) [88] suggesting a treatment-specific phenomenon and a mechanism of acquired resistance.

MSH6 inactivation with loss of MSH6 expression has been observed in patients with grade III and IV posttreatment recurrences [89, 90] indicating that treatment can cause inactivating somatic mutations in subclones of cancer cells which then become TMZ tolerant and lead to tumor progression instead of apoptosis. Similar correlation between TMZ resistance and MSH6 inactivation has been observed in vitro where reconstituting of MSH6 expression restored TMZ sensitivity [91]. The frequency of mutated MSH6 in recurrent GBM after alkylating chemotherapy is about 26 %, while no occurrence has been found in any GBM prior to chemotherapy [88, 91]. Furthermore, mutated MSH6 is associated with a hypermutation phenotype [88, 91] and has also been seen together with mutations in other MMR genes MSH1 and MLH1 [91] (Fig. 3).

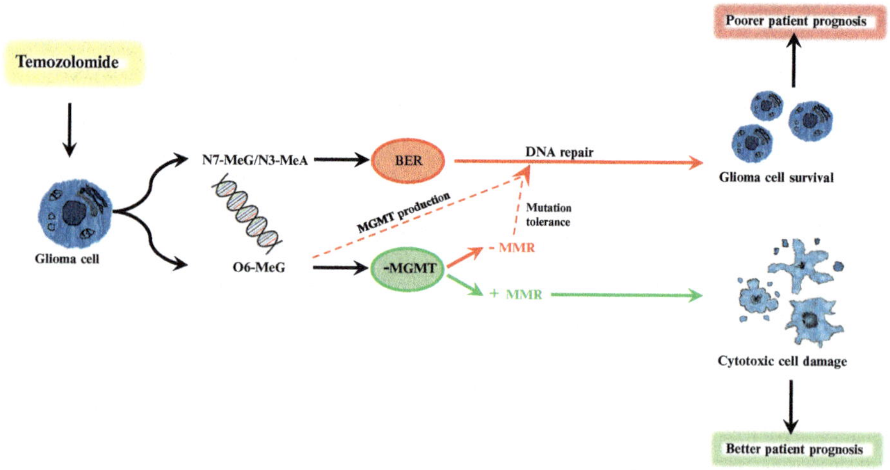

Fig. 3 Escape mechanisms to alkylating chemotherapy. This figure shows possible escape mechanism by which glioma cells escape TMZ-induced cell death depending on the site of the lesion. Overexpression of factors involved in this pathway, e.g., PARP-1 and APNG, potentiates the DNA repair resulting in glioma cell survival and possibly poorer patient prognosis. Figure is based on [70]

In GSC, MMR and its parts appear to be functionally intact. In a series of ten GSC lines, all were MMR proficient [92]. As expected, artificial MSH6 knockdown does reduce the susceptibility of GSC to TMZ [93]. There are though studies showing that GSC have a more efficient DNA repair system (e.g., as monitored by the comet assay) compared to non-stem cells. After DNA damage, GSC stopped proliferation by activation of CDK1/CDK2 checkpoint kinases giving the cells more time to repair the damage. This resulted in an increased radioresistance of GSC when treated with 2.5 Gy under experimental conditions [21]. In which way the better DNA repair system influences the resistance to TMZ is unknown. Given the fact that MMR is required for TMZ-induced strand break, an improved MMR could be associated with an increased TMZ resistance of GSC [48].

The Base Excision Repair Pathway

A well-characterized mechanism of resistance is the inhibition of cell death induced by double-strand breaks. Signal cascades involved include poly(ADP-ribose)-polymerase (PARP) and the base excision repair (BER) pathway, but also mutations in p53 and other genes modulating the apoptotic cascade.

The BER pathway influences the repair of N7-methylguanine (N7-MeG) and N3-methyladenine (N3-MeA) DNA lesions induced by CNU (and to a lesser extent TMZ) by removing bases and thereby securing cell survival. Normally, N7-MeG and N3-MeA do not contribute to the toxicity of TMZ because they were promptly removed by BER. Especially, N3-MeA is highly cytotoxic if it remains uncut by the BER [87]. Within the BER system, APNG is responsible for the excision of the alkylated bases in DNA, and after removing them APNG creates an apurinic/apyrimidinic site (AP-sites). In the following DNA repair process the 5′ (backbone) at the abasic site is hydrolyzed by apurinic-apyrimidinic endonuclease-1 (APE-1) recruited by APNG, resulting in a 5′deoxyribose phosphate (5′dRP) termini and single-strand breaks [94, 95]. DNA polymerase-β fills the gap and removes the abasic remains, which enables the DNA ligase to ligate the nick. The protein XRCC1, a key protein in BER, coordinates this process by working as a scaffold [95–97].

The lack or loss of APNG compromises this strictly coordinated repair process. This can possibly result in accumulated intermediates (like single-strand breaks, 5′dRP, and abasic sites) and eventually in double-strand breaks, chromosome damage, and cell death in cells exposed to different alkylating agents, including agents that almost exclusively produce N3-methyladenine adducts [95, 97–99].

PARP-1 is a key enzyme in BER and is thought to influence sensitivity to TMZ [100] by inhibiting the repair of N3-MeA. However, it can also mediate apoptosis through a mitochondrial pathway that is independent of caspases [101]. Another important player in both the BER pathway and possible mediator in TMZ resistance is apurinic-apyrimidinic endonuclease-1 (APE-1) [102, 103].

There are no detailed studies on the BER pathway and GSC. A recent study using xenograft models of GBM and thereby a GSC-like system confirmed that expression of APNG results in increased TMZ resistance compared to cell lines not expressing APNG [104]. Conversely, the loss of APNG results in an enhanced TMZ sensitivity [104, 105] supporting the concept that BER inhibition could be a promising way to overcome TMZ resistance also of GSC. A recent study systematically investigated the effects of PARP-1 inhibitors on a series of GSC lines. In eight out of ten GSC investigated by Tentori et al. [92], PARP-1 inhibition resulted in an increased susceptibility to TMZ. However, there was no correlation of the effects of PARP-1 treatment with neither PARP-1 expression nor PARP-1 activity.

Multidrug Resistance Proteins

In many different types of cancer, an important mechanism of drug resistance is the presence of multidrug resistance proteins (MDRs), which include ATP-binding cassette transporter proteins, the ABC superfamily (Table 1). MDR contribute to both intrinsic and extrinsic chemoresistance of GBM. MDRs such as breast cancer resistance protein (BCRP1) and multidrug-resistance-associated protein (MRP1) are abundantly expressed in the endothelium of brain capillaries creating a mechanism of extrinsic resistance [106] by avoiding the penetration of the chemotherapeutic into the brain. This mechanism protects tumor cells invading the brain parenchyma while bulk tumor cells are exposed to plasma concentration due to the disrupted blood-brain barrier in the tumor core.

In vitro, TMZ has been shown to be a target of MDR1, and glioma cells positive for MDR1 displayed significantly better viability after exposure to TMZ [107]. Furthermore, the protein of MDR1, P-glycoprotein (P-gp), was reported to contribute to intrinsic chemoresistance in a subgroup of gliomas [108]. High expression of MRP1 has also been associated to chemoresistance in vitro playing a constitutive role in the intrinsic chemoresistance of gliomas [108], and inhibition of MRP1 was reported to increase chemosensitivity [108, 109].

In glioma cell lines with acquired resistance to the alkylating agent BCNU, MRP1 was found to be upregulated and furthermore associated with the epithelial-to-mesenchymal transition (EMT) indicating that development of chemoresistance involves alterations in morphological, molecular, and functional aspects [110]. EMT-like changes have also been seen after chronic exposure to TMZ [111].

MDR proteins have also been investigated in GSC. In experimental settings, a subgroup of MDR proteins, mainly BCRP1/ABCG2, are responsible for the efflux of Hoechst 33342 dye. If a subgroup of cells express BCRP1/ABCG2, these cells become able to efflux this lipophilic and membrane-permeable dye. While other subgroups cannot remove the dye and are stained, the BCRP1/ABCG2-expressing cells appear as distinct, unstained population in cytometric analysis—the "side population." This side population is associated in stem cell properties and their relevance is best characterized a.o. in the bone marrow. Other genes of the MDR family, like MDR-1, do not contribute to the efflux of Hoechst 33342.

In GSC, results of a pilot study suggested that $CD133^+$ GSC express mRNA BCRP1/ABCG2 to a higher degree than $CD133^-$ cells suggesting an overexpression of BRCP1/ABCG2 in GSC. Other papers report that TMZ treatment results in an increase of the proportion of cells within the side population, which is in line with the concept of TMZ selecting for GSC [38, 53]. This suggests that GSC are more resistant to TMZ due to the expression of this MDR gene. However, the role of the side population in GSC and BCRP1/ABCG2 expression is possibly more complicated, a.o. because BCRP1/ABCG2 does not transport TMZ in murine GBM cells [53]. In summary, there is robust data that TMZ increase the proportion of cells in the side population. If the function of MDR proteins as detoxifying proteins is also the cause of the increase of the "side population" or if this increase mirrors an epiphenomenon remains to be clarified.

Other Mediators of TMZ Resistance

The final result of the TMZ-induced DNA damage is apoptosis. Dysregulation of genes and proteins related to apoptosis can therefore play a pivotal role in the resistance of glioma cells to chemotherapy. In GBM, mutations of the p53 are common and crucial [112]. p53 is the major sensor of DNA damage and initiates cell cycle arrest and apoptosis after DNA damage. Inactive p53 has been reported to reduce chemosensitivity to alkylating agents compared to wild-type p53 [86]. However, the p53 system seems to have a more important function in mediating resistance to radiotherapy (Table 1).

MicroRNAs (miRs) have regulatory functions in protein expression [113] and dysregulation of various miRs can cause major shifts in which genes are expressed. Several miRs have been reported to contribute to the acquired chemoresistance seen in GBM both in vitro and in vivo [114, 115], and to regulate expression of MDR components. MiR-21 is considered an important pro-oncogenic factor, i.e., by favoring the antiapoptotic pathway with high expression of Bcl-2 and diminished activity of caspase-3 [114]. Acquired resistance to TMZ may also be coupled to changes in the mitochondrial machinery and the production of reactive oxygen species (ROS) within the glioma cells [116] including metabolic adaptations within the glioma cell population [117]. Components of the antioxidation and detoxification defense system have been correlated with chemoresistance in brain tumor cell lines including glutathione-s-transferase (GST) [118] and aldehyde dehydrogenases (ALDHs) [119].

Biomarkers for Intrinsic Chemoresistance

Several of the abovementioned proteins have already been correlated with malignancy and prognosis. Screening of GBM patients for factors contributing to chemoresistance could therefore potentially help to determine whether the individual patient would benefit from a specific treatment and be part of a personalized medicine approach. Methylation of the MGMT promoter is still the best biomarker for the patient's response to alkylating therapy including TMZ and BCNU [78, 81, 85]. A recent study from Monika Hegi's group now nicely explains why only the MGMT promoter status and not the MGMT expression or the biological MGMT activity in the tumor tissue correlates with the patient's outcome. The explanation implies that GSC have a distinct methylation pattern that is not necessarily shared with other cells within the tumor [84].

Multidrug-Resistant Proteins

The expression level of MDR1 has been shown to positively correlate with glioma malignancy grade. Further, the translated product of MDR1, P-gp, was demonstrated to increase after treatment, and in patients with low-grade glioma overexpression of P-gp was shown to negatively correlate with patient prognosis [106]. Moreover, a single-nucleotide polymorphism in the MDR1 gene, associated with lower gene expression, has been reported to be an independent prognostic factor in GBM patients treated with TMZ resulting in better overall survival. Also multivariate analysis showed that the same MDR1 genotype could be used as an independent predictive factor for the outcome of TMZ therapy [107]. How MDR1 may indicate the response of GSC to TMZ is though unknown. MRP1 and BRCP have also been suggested useful as independent prognostic markers for chemoresistance [106] but a study investigating the promoter methylation failed to predict the patient survival after TMZ therapy [120].

BER Pathway

PARP-1 protein was expressed in the nuclei in tumor tissue from GBM patients, but remained undetectable in normal brain tissue [121]. Studies failed to identify PARP-1 expression as independent prognostic factor [122]. In a study performed on a large number glioma samples ($n = 240$), hypermethylation of the PARP-1 promoter—resulting in reduced expression—was unexpectedly found to correlate positively with malignancy grade and to be associated with shorter progression-free and overall survival [123] possibly because of the underlying CpG island polymethylator phenotype (CIMP). APE-1 has been suggested as a predictive marker in treatment with alkylating agents especially in patients with grade III tumors [103], and it may also be a potential therapeutic target in various cancers [70].

As described above different markers have been found to be biologically significant and important in terms of explaining chemoresistance. However, at the present stage only MGMT is used in the clinical setting.

Mechanisms of Extrinsic Resistance of GSC

The resistance of GSC to TMZ does not only depend on the expression of genes mediating direct resistance to TMZ by detoxifying DNA alkylation. There are a variety of extrinsic factors that substantially contribute to the chemoresistance. One major contributor to chemoresistance, the blood-brain barrier, is not discussed in this paragraph because it protects all tumor cells irrespective of their stem cell state against a multitude of chemotherapeutic drugs.

The susceptibility of GSC is also modified by the metabolic conditions of the microenvironment ("niche") and auto- and paracrine signaling by metabolites and cytokines secreted by tumor cells or invading non-tumor cells like microglia [41, 124]. They either modulate the expression of known mediators of resistance to TMZ or induce resistance by modulating the stem cell state, cell proliferation ("cellular quiescence"), or by yet unknown mechanism.

The Regulation of Intrinsic Resistance by Extrinsic Factors

MGMT is the major mediator of TMZ resistance and the regulation of MGMT expression is therefore of outstanding importance as it can substantially modulate the resistance of GSC. Hypoxia is common in GBM in spite of the extensive vascularization. While cells in the tumor periphery profit from the excellent vascularization of the brain parenchyma, hypoxic areas are typically located in the center of the tumors. This hypoxia is further increased by radiotherapy which may therefore counteract TMZ treatment that is given concomitant to radiotherapy.

Hypoxia is a potent inducer of stemness in GBM cells that depends on HIF-1alpha [125]. HIF-1alpha also regulates MGMT by BMP2 and thereby induces chemoresistance. In line with this concept, the same group identified an increased expression of MGMT around hypoxic areas within tumor samples [126]. Recently, a mTOR-NDRG1 signaling pathway was identified that also resulted in a stabilization of MGMT and thereby chemoresistance [127]. In summary, different groups unequivocally reported an increase of MGMT expression under hypoxia resulting in an increased chemoresistance. This effect appears to be transient, because GSC cultured from cell derived from the tumor core and peripheral parts of the tumor do not consistently differ with respect to their TMZ susceptibility [128].

Concluding Remarks

While GSC are most likely the cause of the invariable tumor recurrence of patients with GBM, the underlying biology of the survival of GSC is less clear. About half of the patients hardly profit from the standard therapy based on TMZ because their

GSC abundantly express of MGMT, resulting in a highly TMZ-resistant phenotype. Since the inhibition of MGMT is associated with a substantially increased toxicity [129] but not clinically feasible, other approaches than alkylating chemotherapy are needed. In GBM not expressing MGMT, the situation is more complicated and it is likely that several different factors including intrinsic and extrinsic factors result in the survival of GSC. The identification of the relevant factors is, however, needed to substantially improve the overall survival of these patients.

References

1. Stupp R, Hegi ME, Mason WP, van den Bent MJ, Taphoorn MJ, Janzer RC, Ludwin SK, Allgeier A, Fisher B, Belanger K, Hau P, Brandes AA, Gijtenbeek J, Marosi C, Vecht CJ, Mokhtari K, Wesseling P, Villa S, Eisenhauer E, Gorlia T, Weller M, Lacombe D, Cairncross JG, Mirimanoff RO. Effects of radiotherapy with concomitant and adjuvant temozolomide versus radiotherapy alone on survival in glioblastoma in a randomised phase III study: 5-year analysis of the EORTC-NCIC trial. Lancet Oncol. 2009;10(5):459–66.
2. Stupp R, Mason WP, van den Bent MJ, Weller M, Fisher B, Taphoorn MJ, Belanger K, Brandes AA, Marosi C, Bogdahn U, Curschmann J, Janzer RC, Ludwin SK, Gorlia T, Allgeier A, Lacombe D, Cairncross JG, Eisenhauer E, Mirimanoff RO. Radiotherapy plus concomitant and adjuvant temozolomide for glioblastoma. N Engl J Med. 2005;352(10):987–96.
3. Ohgaki H, Kleihues P. Epidemiology and etiology of gliomas. Acta Neuropathol (Berl). 2005;109(1):93–108.
4. Louis DN, Ohgaki H, Wiestler OD, Cavenee WK, Burger PC, Jouvet A, Scheithauer BW, Kleihues P. The 2007 WHO classification of tumours of the central nervous system. Acta Neuropathol (Berl). 2007;114(2):97–109.
5. Curran WJ, Scott CB, Horton J, Nelson JS, Weinstein AS, Fischbach AJ, Chang CH, Rotman M, Asbell SO, Krisch RE, Nelson DF. Recursive Partitioning Analysis of Prognostic Factors in 3 Radiation-Therapy Oncology Group Malignant Glioma Trials. J Natl Cancer Inst. 1993;85(9):704–10. doi:10.1093/jnci/85.9.704.
6. Kleihues P, Ohgaki H. Primary and secondary glioblastomas: from concept to clinical diagnosis. Neuro Oncol. 1999;1(1):44–51.
7. Phillips HS, Kharbanda S, Chen R, Forrest WF, Soriano RH, Wu TD, Misra A, Nigro JM, Colman H, Soroceanu L, Williams PM, Modrusan Z, Feuerstein BG, Aldape K. Molecular subclasses of high-grade glioma predict prognosis, delineate a pattern of disease progression, and resemble stages in neurogenesis. Cancer Cell. 2006;9(3):157–73.
8. Verhaak RG, Hoadley KA, Purdom E, Wang V, Qi Y, Wilkerson MD, Miller CR, Ding L, Golub T, Mesirov JP, Alexe G, Lawrence M, O'Kelly M, Tamayo P, Weir BA, Gabriel S, Winckler W, Gupta S, Jakkula L, Feiler HS, Hodgson JG, James CD, Sarkaria JN, Brennan C, Kahn A, Spellman PT, Wilson RK, Speed TP, Gray JW, Meyerson M, Getz G, Perou CM, Hayes DN. Integrated genomic analysis identifies clinically relevant subtypes of glioblastoma characterized by abnormalities in PDGFRA, IDH1, EGFR, and NF1. Cancer Cell. 2010;17(1):98–110.
9. Huse JT, Phillips HS, Brennan CW. Molecular subclassification of diffuse gliomas: seeing order in the chaos. Glia. 2011;59(8):1190–9. doi:10.1002/glia.21165.
10. Varghese M, Olstorn H, Sandberg C, Vik-Mo EO, Noordhuis P, Nister M, Berg-Johnsen J, Moe MC, Langmoen IA. A comparison between stem cells from the adult human brain and from brain tumors. Neurosurgery. 2008;63(6):1022–33.
11. Galli R, Binda E, Orfanelli U, Cipelletti B, Gritti A, De Vitis S, Fiocco R, Foroni C, Dimeco F, Vescovi A. Isolation and characterization of tumorigenic, stem-like neural precursors from human glioblastoma. Cancer Res. 2004;64(19):7011–21.

12. Ignatova TN, Kukekov VG, Laywell ED, Suslov ON, Vrionis FD, Steindler DA. Human cortical glial tumors contain neural stem-like cells expressing astroglial and neuronal markers in vitro. Glia. 2002;39(3):193–206.
13. Singh SK, Clarke ID, Terasaki M, Bonn VE, Hawkins C, Squire J, Dirks PB. Identification of a cancer stem cell in human brain tumors. Cancer Res. 2003;63(18):5821–8.
14. Sanai N, Alvarez-Buylla A, Berger MS. Neural stem cells and the origin of gliomas. N Engl J Med. 2005;353(8):811–22.
15. Bjerkvig R, Tysnes BB, Aboody KS, Najbauer J, Terzis AJ. Opinion: the origin of the cancer stem cell: current controversies and new insights. Nat Rev Cancer. 2005;5(11):899–904.
16. Reya T, Morrison SJ, Clarke MF, Weissman IL. Stem cells, cancer, and cancer stem cells. Nature. 2001;414(6859):105–11.
17. Hamburger AW, Salmon SE. Primary bioassay of human tumor stem cells. Science. 1977;197(4302):461–3.
18. Kreso A, Dick JE. Evolution of the cancer stem cell model. Cell Stem Cell. 2014;14(3): 275–91. doi:10.1016/j.stem.2014.02.006.
19. Luchman HA, Stechishin OD, Dang NH, Blough MD, Chesnelong C, Kelly JJ, Nguyen SA, Chan JA, Weljie AM, Cairncross JG, Weiss S. An in vivo patient-derived model of endogenous IDH1-mutant glioma. Neuro Oncol. 2012;14(2):184–91. doi:10.1093/neuonc/nor207.
20. Pollard SM, Yoshikawa K, Clarke ID, Danovi D, Stricker S, Russell R, Bayani J, Head R, Lee M, Bernstein M, Squire JA, Smith A, Dirks P. Glioma stem cell lines expanded in adherent culture have tumor-specific phenotypes and are suitable for chemical and genetic screens. Cell Stem Cell. 2009;4(6):568–80.
21. Bao S, Wu Q, McLendon RE, Hao Y, Shi Q, Hjelmeland AB, Dewhirst MW, Bigner DD, Rich JN. Glioma stem cells promote radioresistance by preferential activation of the DNA damage response. Nature. 2006;444(7120):756–60.
22. Beier D, Hau P, Proescholdt M, Lohmeier A, Wischhusen J, Oefner PJ, Aigner L, Brawanski A, Bogdahn U, Beier CP. CD133+ and CD133− glioblastoma-derived cancer stem cells show differential growth characteristics and molecular profiles. Cancer Res. 2007;67(9):4010–5.
23. Eramo A, Ricci-Vitiani L, Zeuner A, Pallini R, Lotti F, Sette G, Pilozzi E, Larocca LM, Peschle C, De Maria R. Chemotherapy resistance of glioblastoma stem cells. Cell Death Differ. 2006;13(7):1238–41.
24. Gunther HS, Schmidt NO, Phillips HS, Kemming D, Kharbanda S, Soriano R, Modrusan Z, Meissner H, Westphal M, Lamszus K. Glioblastoma-derived stem cell-enriched cultures form distinct subgroups according to molecular and phenotypic criteria. Oncogene. 2008;27(20):2897–909.
25. Hemmati HD, Nakano I, Lazareff JA, Masterman-Smith M, Geschwind DH, Bronner-Fraser M, Kornblum HI. Cancerous stem cells can arise from pediatric brain tumors. Proc Natl Acad Sci U S A. 2003;100(25):15178–83.
26. Yuan X, Curtin J, Xiong Y, Liu G, Waschsmann-Hogiu S, Farkas DL, Black KL, Yu JS. Isolation of cancer stem cells from adult glioblastoma multiforme. Oncogene. 2004;23(58): 9392–400.
27. Lee J, Kotliarova S, Kotliarov Y, Li A, Su Q, Donin NM, Pastorino S, Purow BW, Christopher N, Zhang W, Park JK, Fine HA. Tumor stem cells derived from glioblastomas cultured in bFGF and EGF more closely mirror the phenotype and genotype of primary tumors than do serum-cultured cell lines. Cancer Cell. 2006;9(5):391–403.
28. Friedmann-Morvinski D, Bushong EA, Ke E, Soda Y, Marumoto T, Singer O, Ellisman MH, Verma IM. Dedifferentiation of neurons and astrocytes by oncogenes can induce gliomas in mice. Science. 2012;338(6110):1080–4. doi:10.1126/science.1226929.
29. Holland EC, Celestino J, Dai C, Schaefer L, Sawaya RE, Fuller GN. Combined activation of Ras and Akt in neural progenitors induces glioblastoma formation in mice. Nat Genet. 2000;25(1):55–7.
30. Zhu Y, Guignard F, Zhao D, Liu L, Burns DK, Mason RP, Messing A, Parada LF. Early inactivation of p53 tumor suppressor gene cooperating with NF1 loss induces malignant astrocytoma. Cancer Cell. 2005;8(2):119–30.

31. Alcantara Llaguno S, Chen J, Kwon CH, Jackson EL, Li Y, Burns DK, Alvarez-Buylla A, Parada LF. Malignant astrocytomas originate from neural stem/progenitor cells in a somatic tumor suppressor mouse model. Cancer Cell. 2009;15(1):45–56.
32. Zheng H, Ying H, Yan H, Kimmelman AC, Hiller DJ, Chen AJ, Perry SR, Tonon G, Chu GC, Ding Z, Stommel JM, Dunn KL, Wiedemeyer R, You MJ, Brennan C, Wang YA, Ligon KL, Wong WH, Chin L, DePinho RA. p53 and Pten control neural and glioma stem/progenitor cell renewal and differentiation. Nature. 2008;455(7216):1129–33.
33. Jang T, Litofsky NS, Smith TW, Ross AH, Recht LD. Aberrant nestin expression during ethylnitrosourea-(ENU)-induced neurocarcinogenesis. Neurobiol Dis. 2004;15(3):544–52.
34. Alvarez-Buylla A, Lim DA. For the long run: maintaining germinal niches in the adult brain. Neuron. 2004;41(5):683–6.
35. Zhao CM, Deng W, Gage FH. Mechanisms and functional implications of adult neurogenesis. Cell. 2008;132(4):645–60. doi:10.1016/j.cell.2008.01.033.
36. Glass R, Synowitz M, Kronenberg G, Walzlein JH, Markovic DS, Wang LP, Gast D, Kiwit JRR, Kempermann G, Kettenmann H. Glioblastoma-induced attraction of endogenous neural precursor cells is associated with improved survival. J Neurosci. 2005;25(10):2637–46. doi:10.1523/jneurosci. 5118-04.2005.
37. Doetsch F, Petreanu L, Caille I, Garcia-Verdugo JM, Alvarez-Buylla A. EGF converts transit-amplifying neurogenic precursors in the adult brain into multipotent stem cells. Neuron. 2002;36(6):1021–34.
38. Chua C, Zaiden N, Chong KH, See SJ, Wong MC, Ang BT, Tang C. Characterization of a side population of astrocytoma cells in response to temozolomide. J Neurosurg. 2008;109(5): 856–66.
39. Wan F, Zhang S, Xie R, Gao B, Campos B, Herold-Mende C, Lei T. The utility and limitations of neurosphere assay, CD133 immunophenotyping and side population assay in glioma stem cells research. Brain Pathol. 2010;9999(999A).
40. Chen R, Nishimura MC, Bumbaca SM, Kharbanda S, Forrest WF, Kasman IM, Greve JM, Soriano RH, Gilmour LL, Rivers CS, Modrusan Z, Nacu S, Guerrero S, Edgar KA, Wallin JJ, Lamszus K, Westphal M, Heim S, James CD, VandenBerg SR, Costello JF, Moorefield S, Cowdrey CJ, Prados M, Phillips HS. A hierarchy of self-renewing tumor-initiating cell types in glioblastoma. Cancer Cell. 2010;17(4):362–75.
41. Li Z, Bao S, Wu Q, Wang H, Eyler C, Sathornsumetee S, Shi Q, Cao Y, Lathia J, McLendon RE, Hjelmeland AB, Rich JN. Hypoxia-inducible factors regulate tumorigenic capacity of glioma stem cells. Cancer Cell. 2009;15(6):501–13.
42. Chen J, Li YJ, Yu TS, McKay RM, Burns DK, Kernie SG, Parada LF. A restricted cell population propagates glioblastoma growth after chemotherapy. Nature. 2012;488(7412):522. doi:10.1038/nature11287.
43. Auffinger B, Tobias AL, Han Y, Lee G, Guo D, Dey M, Lesniak MS, Ahmed AU. Conversion of differentiated cancer cells into cancer stem-like cells in a glioblastoma model after primary chemotherapy. Cell Death Differ. 2014;21(7):1119–31. doi:10.1038/cdd.2014.31.
44. Brennan C, Momota H, Hambardzumyan D, Ozawa T, Tandon A, Pedraza A, Holland E. Glioblastoma subclasses can be defined by activity among signal transduction pathways and associated genomic alterations. PLoS One. 2009;4(11):e7752.
45. Bhat KP, Salazar KL, Balasubramaniyan V, Wani K, Heathcock L, Hollingsworth F, James JD, Gumin J, Diefes KL, Kim SH, Turski A, Azodi Y, Yang Y, Doucette T, Colman H, Sulman EP, Lang FF, Rao G, Copray S, Vaillant BD, Aldape KD. The transcriptional coactivator TAZ regulates mesenchymal differentiation in malignant glioma. Genes Dev. 2011;25(24):2594–609. doi:10.1101/gad.176800.111.
46. Beier F, Beier CP, Aschenbrenner I, Hildebrandt GC, Brummendorf TH, Beier D. Identification of CD133(−)/telomerase(low) progenitor cells in glioblastoma-derived cancer stem cell lines. Cell Mol Neurobiol. 2011;31(3):337–43. doi:10.1007/s10571-010-9627-4.
47. Grant R, Liang BC, Slattery J, Greenberg HS, Junck L. Chemotherapy response criteria in malignant glioma. Neurology. 1997;48(5):1336–40.

48. Beier D, Schulz JB, Beier CP. Chemoresistance of glioblastoma cancer stem cells–much more complex than expected. Mol Cancer. 2011;10:128. doi:10.1186/1476-4598-10-128.
49. Pallini R, Ricci-Vitiani L, Banna GL, Signore M, Lombardi D, Todaro M, Stassi G, Martini M, Maira G, Larocca LM, De Maria R. Cancer stem cell analysis and clinical outcome in patients with glioblastoma multiforme. Clin Cancer Res. 2008;14(24):8205–12.
50. Pallini R, Ricci-Vitiani L, Montano N, Mollinari C, Biffoni M, Cenci T, Pierconti F, Martini M, De Maria R, Larocca LM. Expression of the stem cell marker CD133 in recurrent glioblastoma and its value for prognosis. Cancer. 2011;117(1):162–74. doi:10.1002/cncr.25581.
51. Wang J, Wakeman TP, Lathia JD, Hjelmeland AB, Wang XF, White RR, Rich JN, Sullenger BA. Notch promotes radioresistance of glioma stem cells. Stem Cells. 2010;28(1):17–28. doi:10.1002/stem.261.
52. Beier D, Rohrl S, Pillai DR, Schwarz S, Kunz-Schughart LA, Leukel P, Proescholdt M, Brawanski A, Bogdahn U, Trampe-Kieslich A, Giebel B, Wischhusen J, Reifenberger G, Hau P, Beier CP. Temozolomide preferentially depletes cancer stem cells in glioblastoma. Cancer Res. 2008;68(14):5706–15.
53. Bleau AM, Hambardzumyan D, Ozawa T, Fomchenko EI, Huse JT, Brennan CW, Holland EC. PTEN/PI3K/Akt pathway regulates the side population phenotype and ABCG2 activity in glioma tumor stem-like cells. Cell Stem Cell. 2009;4(3):226–35.
54. Liu GT, Yuan XP, Zeng ZH, Tunici P, Ng HS, Abdulkadir IR, Lu LZ, Irvin D, Black KL, Yu JS. Analysis of gene expression and chemoresistance of CDI33(+) cancer stem cells in glioblastoma. Mol Cancer. 2006;5:12. doi:10.1186/1476-4598-5-67.
55. Ghods AJ, Irvin D, Liu G, Yuan X, Abdulkadir IR, Tunici P, Konda B, Wachsmann-Hogiu S, Black KL, Yu JS. Spheres isolated from 9L gliosarcoma rat cell line possess chemoresistant and aggressive cancer stem-like cells. Stem Cells. 2007;25(7):1645–53.
56. Hong X, Chedid K, Kalkanis SN. Glioblastoma cell line-derived spheres in serum-containing medium versus serum-free medium: a comparison of cancer stem cell properties. Int J Oncol. 2012;41(5):1693–700. doi:10.3892/ijo.2012.1592.
57. Clement V, Sanchez P, de Tribolet N, Radovanovic I, Ruiz i Altaba A. HEDGEHOG-GLI1 signaling regulates human glioma growth, cancer stem cell self-renewal, and tumorigenicity. Curr Biol. 2007;17(2):165–72.
58. Villalva C, Martin-Lanneree S, Cortes U, Dkhissi F, Wager M, Le Corf A, Tourani J, Dusanter-Fourt I, Turhan A, Karayan-Tapon L. STAT3 is essential for the maintenance of neurosphere-initiating tumor cells in patients with glioblastomas: a potential for targeted therapy? Int J Cancer. 2011;128(4):826–38.
59. Happold C, Roth P, Silginer M, Florea AM, Lamszus K, Frei K, Deenen R, Reifenberger G, Weller M. Interferon-beta induces loss of spherogenicity and overcomes therapy resistance of glioblastoma stem cells. Mol Cancer Ther. 2014;13(4):948–61. doi:10.1158/1535-7163.MCT-13-0772.
60. Gilbert CA, Daou MC, Moser RP, Ross AH. Gamma-secretase inhibitors enhance temozolomide treatment of human gliomas by inhibiting neurosphere repopulation and xenograft recurrence. Cancer Res. 2010;70(17):6870–9.
61. Beier D, Schriefer B, Brawanski K, Hau P, Weis J, Schulz JB, Beier CP. Efficacy of clinically relevant temozolomide dosing schemes in glioblastoma cancer stem cell lines. J Neurooncol. 2012;109(1):45–52. doi:10.1007/s11060-012-0878-4.
62. Fouse SD, Nagarajan RP, Nakamura J, James CD, Chang S, Costello JF. Therapeutic response in primary glioblastoma neural stem cells relative to patient-matched non-stem tumor cells. Neuro Oncol. 2011;13:151–2.
63. Meads MB, Hazlehurst LA, Dalton WS. The bone marrow microenvironment as a tumor sanctuary and contributor to drug resistance. Clin Cancer Res. 2008;14(9):2519–26.
64. Adamson C, Kanu OO, Mehta AI, Di CH, Lin NJ, Mattox AK, Bigner DD. Glioblastoma multiforme: a review of where we have been and where we are going. Expert Opin Investig Drugs. 2009;18(8):1061–83. doi:10.1517/13543780903052764.
65. Taal W, Oosterkamp HM, Walenkamp AM, Dubbink HJ, Beerepoot LV, Hanse MC, Buter J, Honkoop AH, Boerman D, de Vos FY, Dinjens WN, Enting RH, Taphoorn MJ, van den

Berkmortel FW, Jansen RL, Brandsma D, Bromberg JE, van Heuvel I, Vernhout RM, van der Holt B, van den Bent MJ. Single-agent bevacizumab or lomustine versus a combination of bevacizumab plus lomustine in patients with recurrent glioblastoma (BELOB trial): a randomised controlled phase 2 trial. Lancet Oncol. 2014;15(9):943–53. doi:10.1016/S1470-2045(14)70314-6.

66. Wick W, Puduvalli VK, Chamberlain MC, van den Bent MJ, Carpentier AF, Cher LM, Mason W, Weller M, Hong S, Musib L, Liepa AM, Thornton DE, Fine HA. Phase III study of enzastaurin compared with lomustine in the treatment of recurrent intracranial glioblastoma. J Clin Oncol. 2010;28(7):1168–74. doi:10.1200/JCO.2009.23.2595.
67. Westphal M, Hilt DC, Bortey E, Delavault P, Olivares R, Warnke PC, Whittle IR, Jaaskelainen J, Ram Z. A phase 3 trial of local chemotherapy with biodegradable carmustine (BCNU) wafers (Gliadel wafers) in patients with primary malignant glioma. Neuro Oncol. 2003;5(2):79–88. doi:10.1215/S1522-8517-02-00023-6.
68. Batista LF, Roos WP, Christmann M, Menck CF, Kaina B. Differential sensitivity of malignant glioma cells to methylating and chloroethylating anticancer drugs: p53 determines the switch by regulating xpc, ddb2, and DNA double-strand breaks. Cancer Res. 2007;67(24): 11886–95.
69. Hermisson M, Klumpp A, Wick W, Wischhusen J, Nagel G, Roos W, Kaina B, Weller M. O-methylguanine DNA methyltransferase and p53 status predict temozolomide sensitivity in human malignant glioma cells. J Neurochem. 2005.
70. Roos WP, Batista LF, Naumann SC, Wick W, Weller M, Menck CF, Kaina B. Apoptosis in malignant glioma cells triggered by the temozolomide-induced DNA lesion O6-methylguanine. Oncogene. 2007;26(2):186–97.
71. Margison GP, Santibanez-Koref MF. O6-alkylguanine-DNA alkyltransferase: role in carcinogenesis and chemotherapy. BioEssays. 2002;24(3):255–66. doi:10.1002/bies.10063.
72. Day III RS, Ziolkowski CH, Scudiero DA, Meyer SA, Lubiniecki AS, Girardi AJ, Galloway SM, Bynum GD. Defective repair of alkylated DNA by human tumour and SV40-transformed human cell strains. Nature. 1980;288(5792):724–7.
73. Nakagawachi T, Soejima H, Urano T, Zhao W, Higashimoto K, Satoh Y, Matsukura S, Kudo S, Kitajima Y, Harada H, Furukawa K, Matsuzaki H, Emi M, Nakabeppu Y, Miyazaki K, Sekiguchi M, Mukai T. Silencing effect of CpG island hypermethylation and histone modifications on O6-methylguanine-DNA methyltransferase (MGMT) gene expression in human cancer. Oncogene. 2003;22(55):8835–44. doi:10.1038/sj.onc.1207183.
74. von Deimling A, Korshunov A, Hartmann C. The next generation of glioma biomarkers: MGMT methylation, BRAF fusions and IDH1 mutations. Brain Pathol (Zurich, Switzerland). 2011;21(1):74–87. doi:10.1111/j.1750-3639.2010.00454.x.
75. Pegg AE, Fang Q, Loktionova NA. Human variants of O6-alkylguanine-DNA alkyltransferase. DNA Repair. 2007;6(8):1071–8. doi:10.1016/j.dnarep.2007.03.012.
76. Sharma S, Salehi F, Scheithauer BW, Rotondo F, Syro LV, Kovacs K. Role of MGMT in tumor development, progression, diagnosis, treatment and prognosis. Anticancer Res. 2009; 29(10):3759–68.
77. Weller M, Stupp R, Reifenberger G, Brandes AA, van den Bent MJ, Wick W, Hegi ME. MGMT promoter methylation in malignant gliomas: ready for personalized medicine? Nat Rev Neurol. 2010;6(1):39–51. doi:10.1038/nrneurol.2009.197.
78. Hegi ME, Diserens AC, Gorlia T, Hamou MF, de Tribolet N, Weller M, Kros JM, Hainfellner JA, Mason W, Mariani L, Bromberg JE, Hau P, Mirimanoff RO, Cairncross JG, Janzer RC, Stupp R. MGMT gene silencing and benefit from temozolomide in glioblastoma. N Engl J Med. 2005;352(10):997–1003.
79. Gerson SL. MGMT: its role in cancer aetiology and cancer therapeutics. Nat Rev Cancer. 2004;4(4):296–307. doi:10.1038/nrc1319.
80. Gong X, Schwartz PH, Linskey ME, Bota DA. Neural stem/progenitors and glioma stem-like cells have differential sensitivity to chemotherapy. Neurology. 2011;76(13):1126–34. doi:10.1212/WNL.0b013e318212a89f.

81. Riemenschneider MJ, Jeuken JW, Wesseling P, Reifenberger G. Molecular diagnostics of gliomas: state of the art. Acta Neuropathol. 2010;120(5):567–84. doi:10.1007/s00401-010-0736-4.
82. Preusser M, Janzer RC, Felsberg J, Reifenberger G, Hamou MF, Diserens AC, Stupp R, Gorlia T, Marosi C, Heinzl H, Hainfellner JA, Hegi M. Anti-O6-methylguanine-methyltransferase (MGMT) immunohistochemistry in glioblastoma multiforme: observer variability and lack of association with patient survival impede its use as clinical biomarker. Brain Pathol. 2008;18(4):520–32. doi:10.1111/j.1750-3639.2008.00153.x.
83. Christmann M, Nagel G, Horn S, Krahn U, Wiewrodt D, Sommer C, Kaina B. MGMT activity, promoter methylation and immunohistochemistry of pretreatment and recurrent malignant gliomas: a comparative study on astrocytoma and glioblastoma. Int J Cancer. 2010;127(9):2106–18. doi:10.1002/ijc.25229.
84. Sciuscio D, Diserens AC, van Dommelen K, Martinet D, Jones G, Janzer RC, Pollo C, Hamou MF, Kaina B, Stupp R, Levivier M, Hegi ME. Extent and patterns of MGMT promoter methylation in glioblastoma- and respective glioblastoma-derived spheres. Clin Cancer Res. 2011;17(2):255–66. doi:10.1158/1078-0432.ccr-10-1931.
85. Esteller M, Garcia-Foncillas J, Andion E, Goodman SN, Hidalgo OF, Vanaclocha V, Baylin SB, Herman JG. Inactivation of the DNA-repair gene MGMT and the clinical response of gliomas to alkylating agents. N Engl J Med. 2000;343(19):1350–4.
86. Sarkaria JN, Kitange GJ, James CD, Plummer R, Calvert H, Weller M, Wick W. Mechanisms of chemoresistance to alkylating agents in malignant glioma. Clin Cancer Res. 2008;14(10):2900–8. doi:10.1158/1078-0432.ccr-07-1719.
87. Zhang J, Stevens MF, Bradshaw TD. Temozolomide: mechanisms of action, repair and resistance. Curr Mol Pharmacol. 2012;5(1):102–14.
88. Cancer Genome Atlas Research Network. Comprehensive genomic characterization defines human glioblastoma genes and core pathways. Nature. 2008;455(7216):1061–8. doi:10.1038/nature07385.
89. Cahill DP, Levine KK, Betensky RA, Codd PJ, Romany CA, Reavie LB, Batchelor TT, Futreal PA, Stratton MR, Curry WT, Iafrate AJ, Louis DN. Loss of the mismatch repair protein MSH6 in human glioblastomas is associated with tumor progression during temozolomide treatment. Clin Cancer Res. 2007;13(7):2038–45.
90. Hunter C, Smith R, Cahill DP, Stephens P, Stevens C, Teague J, Greenman C, Edkins S, Bignell G, Davies H, O'Meara S, Parker A, Avis T, Barthorpe S, Brackenbury L, Buck G, Butler A, Clements J, Cole J, Dicks E, Forbes S, Gorton M, Gray K, Halliday K, Harrison R, Hills K, Hinton J, Jenkinson A, Jones D, Kosmidou V, Laman R, Lugg R, Menzies A, Perry J, Petty R, Raine K, Richardson D, Shepherd R, Small A, Solomon H, Tofts C, Varian J, West S, Widaa S, Yates A, Easton DF, Riggins G, Roy JE, Levine KK, Mueller W, Batchelor TT, Louis DN, Stratton MR, Futreal PA, Wooster R. A hypermutation phenotype and somatic MSH6 mutations in recurrent human malignant gliomas after alkylator chemotherapy. Cancer Res. 2006;66(8):3987–91.
91. Yip S, Miao J, Cahill DP, Iafrate AJ, Aldape K, Nutt CL, Louis DN. MSH6 mutations arise in glioblastomas during temozolomide therapy and mediate temozolomide resistance. Clin Cancer Res. 2009;15(14):4622–9.
92. Tentori L, Ricci-Vitiani L, Muzi A, Ciccarone F, Pelacchi F, Calabrese R, Runci D, Pallini R, Caiafa P, Graziani G. Pharmacological inhibition of poly(ADP-ribose) polymerase-1 modulates resistance of human glioblastoma stem cells to temozolomide. BMC Cancer. 2014; 14(1):151.
93. Kanai R, Rabkin SD, Yip S, Sgubin D, Zaupa CM, Hirose Y, Louis DN, Wakimoto H, Martuza RL. Oncolytic virus-mediated manipulation of DNA damage responses: synergy with chemotherapy in killing glioblastoma stem cells. J Natl Cancer Inst. 2012;104(1):42–55. doi:10.1093/jnci/djr509.
94. Silber JR, Bobola MS, Kolstoe DD, Blank A, Chamberlain MC. Repair of 3-methyladenine and abasic sites by base excision repair mediates glioblastoma resistance to temozolomide. Front Oncol. 2012;2:176. doi:10.3389/fonc.2012.00176.

95. Fu D, Calvo JA, Samson LD. Balancing repair and tolerance of DNA damage caused by alkylating agents. Nat Rev Cancer. 2012;12(2):104–20. doi:10.1038/nrc3185.
96. Smith SA, Engelward BP. In vivo repair of methylation damage in Aag 3-methyladenine DNA glycosylase null mouse cells. Nucleic Acids Res. 2000;28(17):3294–300. doi:10.1093/nar/28.17.3294.
97. Bobola MS, Silber JR, Ellenbogen RG, Geyer JR, Blank A, Goff RD. O6-methylguanine-DNA methyltransferase, O6-benzylguanine, and resistance to clinical alkylators in pediatric primary brain tumor cell lines. Clin Cancer Res. 2005;11(7):2747–55.
98. Hang B, Singer B, Margison GP, Elder RH. Targeted deletion of alkylpurine-DNA-N-glycosylase in mice eliminates repair of 1, N-6-ethenoadenine and hypoxanthine but not of 3, N-4-ethenocytosine or 8-oxoguanine. Proc Natl Acad Sci U S A. 1997;94(24):12869–74. doi:10.1073/pnas.94.24.12869.
99. Engelward BP, Dreslin A, Christensen J, Huszar D, Kurahara C, Samson L. Repair-deficient 3-methyladenine DNA glycosylase homozygous mutant mouse cells have increased sensitivity to alkylation-induced chromosome damage and cell killing. EMBO J. 1996;15(4):945–52.
100. Tentori L, Portarena I, Torino F, Scerrati M, Navarra P, Graziani G. Poly(ADP-ribose) polymerase inhibitor increases growth inhibition and reduces G(2)/M cell accumulation induced by temozolomide in malignant glioma cells. Glia. 2002;40(1):44–54. doi:10.1002/glia.10113.
101. Yu SW, Wang H, Poitras MF, Coombs C, Bowers WJ, Federoff HJ, Poirier GG, Dawson TM, Dawson VL. Mediation of poly(ADP-ribose) polymerase-1-dependent cell death by apoptosis-inducing factor. Science (New York, NY). 2002;297(5579):259–63. doi:10.1126/science.1072221.
102. Silber JR, Bobola MS, Blank A, Schoeler KD, Haroldson PD, Huynh MB, Kolstoe DD. The apurinic/apyrimidinic endonuclease activity of Ape1/Ref-1 contributes to human glioma cell resistance to alkylating agents and is elevated by oxidative stress. Clin Cancer Res. 2002;8(9):3008–18.
103. Bobola MS, Emond MJ, Blank A, Meade EH, Kolstoe DD, Berger MS, Rostomily RC, Silbergeld DL, Spence AM, Silber JR. Apurinic endonuclease activity in adult gliomas and time to tumor progression after alkylating agent-based chemotherapy and after radiotherapy. Clin Cancer Res. 2004;10(23):7875–83. doi:10.1158/1078-0432.ccr-04-1161.
104. Agnihotri S, Gajadhar A, Ternamian C, Gorlia T, Diefes K, Mischel P, Kelly J, McGown G, Thorncroft M, Carlson B, Sarkaria J, Margison G, Aldape K, Hawkins C, Hegi M, Guha A. Alkylpurine-DNA-N-glycosylase confers resistance to temozolomide in xenograft models of glioblastoma multiforme and is associated with poor survival in patients. J Clin Invest. 2012;122:253–66.
105. Agnihotri S, Wolf A, Munoz DM, Smith CJ, Gajadhar A, Restrepo A, Clarke ID, Fuller GN, Kesari S, Dirks PB, McGlade CJ, Stanford WL, Aldape K, Mischel PS, Hawkins C, Guha A. A GATA4-regulated tumor suppressor network represses formation of malignant human astrocytomas. J Exp Med. 2011;208(4):689–702. doi:10.1084/jem.20102099.
106. Lu C, Shervington A. Chemoresistance in gliomas. Mol Cell Biochem. 2008;312(1–2):71–80. doi:10.1007/s11010-008-9722-8.
107. Schaich M, Kestel L, Pfirrmann M, Robel K, Illmer T, Kramer M, Dill C, Ehninger G, Schackert G, Krex D. A MDR1 (ABCB1) gene single nucleotide polymorphism predicts outcome of temozolomide treatment in glioblastoma patients. Ann Oncol. 2009;20(1):175–81.
108. Spiegl-Kreinecker S, Buchroithner J, Elbling L, Steiner E, Wurm G, Bodenteich A, Fischer J, Micksche M, Berger W. Expression and functional activity of the ABC-transporter proteins P-glycoprotein and multidrug-resistance protein 1 in human brain tumor cells and astrocytes. J Neurooncol. 2002;57(1):27–36.
109. Peignan L, Garrido W, Segura R, Melo R, Rojas D, Carcamo JG, San Martin R, Quezada C. Combined use of anticancer drugs and an inhibitor of multiple drug resistance-associated protein-1 increases sensitivity and decreases survival of glioblastoma multiforme cells in vitro. Neurochem Res. 2011;36(8):1397–406. doi:10.1007/s11064-011-0464-8.

110. Yan YR, Xie Q, Li F, Zhang Y, Ma JW, Xie SM, Li HY, Zhong XY. Epithelial-to-mesenchymal transition is involved in BCNU resistance in human glioma cells. Neuropathology. 2013;34(2):128–34. doi:10.1111/neup.12062.
111. Sun S, Wong TS, Zhang XQ, Pu JK, Lee NP, Day PJ, Ng GK, Lui WM, Leung GK. Protein alterations associated with temozolomide resistance in subclones of human glioblastoma cell lines. J Neurooncol. 2012;107(1):89–100. doi:10.1007/s11060-011-0729-8.
112. Wang X, Li C, Chen Y, Hao Y, Zhou W, Chen C, Yu Z. Hypoxia enhances CXCR4 expression favoring microglia migration via HIF-1alpha activation. Biochem Biophys Res Commun. 2008;371(2):283–8. doi:10.1016/j.bbrc.2008.04.055.
113. Hermansen SK, Kristensen BW. MicroRNA biomarkers in glioblastoma. J Neurooncol. 2013;114(1):13–23. doi:10.1007/s11060-013-1155-x.
114. Koshkin PA, Chistiakov DA, Chekhonin VP. Role of microRNAs in mechanisms of glioblastoma resistance to radio- and chemotherapy. Biochem Biokhim. 2013;78(4):325–34. doi:10.1134/s0006297913040019.
115. Ujifuku K, Mitsutake N, Takakura S, Matsuse M, Saenko V, Suzuki K, Hayashi K, Matsuo T, Kamada K, Nagata I, Yamashita S. miR-195, miR-455-3p and miR-10a(*) are implicated in acquired temozolomide resistance in glioblastoma multiforme cells. Cancer Lett. 2010;296(2):241–8. doi:10.1016/j.canlet.2010.04.013.
116. Oliva CR, Moellering DR, Gillespie GY, Griguer CE. Acquisition of chemoresistance in gliomas is associated with increased mitochondrial coupling and decreased ROS production. PLoS One. 2011;6(9):e24665. doi:10.1371/journal.pone.0024665.
117. Ye F, Zhang Y, Liu Y, Yamada K, Tso JL, Menjivar JC, Tian JY, Yong WH, Schaue D, Mischel PS, Cloughesy TF, Nelson SF, Liau LM, McBride W, Tso CL. Protective properties of radio-chemoresistant glioblastoma stem cell clones are associated with metabolic adaptation to reduced glucose dependence. PLoS One. 2013;8(11):e80397. doi:10.1371/journal.pone.0080397.
118. Mousseau M, Chauvin C, Nissou MF, Chaffanet M, Plantaz D, Pasquier B, Schaerer R, Benabid A. A study of the expression of four chemoresistance-related genes in human primary and metastatic brain tumours. Eur J Cancer (Oxford, Engl 1990). 1993;29a(5):753–9.
119. Schafer A, Teufel J, Ringel F, Bettstetter M, Hoepner I, Rasper M, Gempt J, Koeritzer J, Schmidt-Graf F, Meyer B, Beier CP, Schlegel J. Aldehyde dehydrogenase 1A1–a new mediator of resistance to temozolomide in glioblastoma. Neuro Oncol. 2012;14(12):1452–64. doi:10.1093/neuonc/nos270.
120. Oberstadt MC, Bien-Moller S, Weitmann K, Herzog S, Hentschel K, Rimmbach C, Vogelgesang S, Balz E, Fink M, Michael H, Zeden JP, Bruckmuller H, Werk AN, Cascorbi I, Hoffmann W, Rosskopf D, Schroeder HW, Kroemer HK. Epigenetic modulation of the drug resistance genes MGMT, ABCB1 and ABCG2 in glioblastoma multiforme. BMC Cancer. 2013;13:617. doi:10.1186/1471-2407-13-617.
121. Galia A, Calogero AE, Condorelli R, Fraggetta F, La Corte A, Ridolfo F, Bosco P, Castiglione R, Salemi M. PARP-1 protein expression in glioblastoma multiforme. Eur J Histochem. 2012;56(1):e9. doi:10.4081/ejh.2012.e9.
122. Kase M, Vardja M, Lipping A, Asser T, Jaal J. Impact of PARP-1 and DNA-PK expression on survival in patients with glioblastoma multiforme. Radiother Oncol. 2011;101(1):127–31. doi:10.1016/j.radonc.2011.06.024.
123. Zhang L, Wang M, Wang W, Mo J. Incidence and prognostic value of multiple gene promoter methylations in gliomas. J Neurooncol. 2014;116(2):349–56. doi:10.1007/s11060-013-1301-5.
124. Bhat KP, Balasubramaniyan V, Vaillant B, Ezhilarasan R, Hummelink K, Hollingsworth F, Wani K, Heathcock L, James JD, Goodman LD, Conroy S, Long L, Lelic N, Wang S, Gumin J, Raj D, Kodama Y, Raghunathan A, Olar A, Joshi K, Pelloski CE, Heimberger A, Kim SH, Cahill DP, Rao G, Den Dunnen WF, Boddeke HW, Phillips HS, Nakano I, Lang FF, Colman H, Sulman EP, Aldape K. Mesenchymal differentiation mediated by NF-kappaB promotes radiation resistance in glioblastoma. Cancer Cell. 2013;24(3):331–46. doi:10.1016/j.ccr.2013.08.001.

125. Pistollato F, Abbadi S, Rampazzo E, Persano L, Della Puppa A, Frasson C, Sarto E, Scienza R, D'Avella D, Basso G. Intratumoral hypoxic gradient drives stem cells distribution and MGMT expression in glioblastoma. Stem Cells. 2010;28(5):851–62. doi:10.1002/stem.415.
126. Persano L, Pistollato F, Rampazzo E, Della Puppa A, Abbadi S, Frasson C, Volpin F, Indraccolo S, Scienza R, Basso G. BMP2 sensitizes glioblastoma stem-like cells to Temozolomide by affecting HIF-1 alpha stability and MGMT expression. Cell Death Dis. 2012;3:e412. doi:10.1038/cddis.2012.153.
127. Weiler M, Blaes J, Pusch S, Sahm F, Czabanka M, Luger S, Bunse L, Solecki G, Eichwald V, Jugold M, Hodecker S, Osswald M, Meisner C, Hielscher T, Rubmann P, Pfenning PN, Ronellenfitsch M, Kempf T, Schnolzer M, Abdollahi A, Lang F, Bendszus M, von Deimling A, Winkler F, Weller M, Vajkoczy P, Platten M, Wick W. mTOR target NDRG1 confers MGMT-dependent resistance to alkylating chemotherapy. Proc Natl Acad Sci U S A. 2014;111(1):409–14. doi:10.1073/pnas.1314469111.
128. Glas M, Rath BH, Simon M, Reinartz R, Schramme A, Trageser D, Eisenreich R, Leinhaas A, Keller M, Schildhaus HU, Garbe S, Steinfarz B, Pietsch T, Steindler DA, Schramm J, Herrlinger U, Brustle O, Scheffler B. Residual tumor cells are unique cellular targets in glioblastoma. Ann Neurol. 2010;68(2):264–9. doi:10.1002/ana.22036.
129. Quinn JA, Jiang SX, Reardon DA, Desjardins A, Vredenburgh JJ, Rich JN, Gururangan S, Friedman AH, Bigner DD, Sampson JH, McLendon RE, Herndon JE, Walker A, Friedman HS. Phase II trial of temozolomide plus O-6-benzylguanine in adults with recurrent, temozolomide-resistant malignant glioma. J Clin Oncol. 2009;27(8):1262–7.

Immunobiology and Immunotherapeutic Targeting of Glioma Stem Cells

Mecca Madany, Tom M. Thomas, Lincoln Edwards, and John S. Yu

Abstract For decades human brain tumors have confounded our efforts to effectively manage and treat patients. In adults, glioblastoma multiforme is the most common malignant brain tumor with a patient survival of just over 14 months. In children, brain tumors are the leading cause of solid tumor cancer death and gliomas account for one-fifth of all childhood cancers. Despite advances in conventional treatments such as surgical resection, radiotherapy, and systemic chemotherapy, the incidence and mortality rates for gliomas have essentially stayed the same. Furthermore, research efforts into novel therapeutics that initially appeared promising have yet to show a marked benefit. A shocking and somewhat disturbing view is that investigators and clinicians may have been targeting the wrong cells, resulting in the appearance of the removal or eradication of patient gliomas only to have brain cancer recurrence. Here we review research progress in immunotherapy as it pertains to glioma treatment and how it can and is being adapted to target glioma stem cells (GSCs) as a means of dealing with this potential paradigm.

Keywords Glioma stem cells • Glioblastoma • Cancer stem cells • Immunotherapy • Cancer immunology

M. Madany • T.M. Thomas
Maxine Dunitz Neurosurgical Institute, Cedars-Sinai Medical Center,
127 S. San Vicente Blvd, Los Angeles, CA 90048, USA

Graduate Program in Biological Science & Translational Medicine, Cedars-Sinai Medical Center, Los Angeles, CA, USA

L. Edwards, Ph.D. • J.S. Yu, M.D. (✉)
Maxine Dunitz Neurosurgical Institute, Cedars-Sinai Medical Center,
127 S. San Vicente Blvd, Los Angeles, CA 90048, USA
e-mail: John.Yu@cshs.org

M. Ehtesham (ed.), *Stem Cell Biology in Neoplasms of the Central Nervous System*, Advances in Experimental Medicine and Biology 853,
DOI 10.1007/978-3-319-16537-0_8

Introduction

Why Therapy Has Failed

The cancer stem cell (CSC) theory posits that the failure of patients inflicted with brain cancer, particularly high-grade gliomas such as glioblastoma multiformes (GBMs), is due to a subpopulation of cells with stem cell-like features termed GSCs that are responsible for the malignant behavior of GBM and for treatment failure [1]. This is evidenced by the fact that GSCs have been shown to be radiation resistant [2], capable of initiating tumor growth, and can transform into additional cell types that support or maintain and most likely enhance tumor progression [3–6]. Further thwarting our efforts to treat patients effectively, GSCs can activate DNA repair mechanisms, have self-renewal properties (being able to produce more of itself), and have drug transporters allowing them to withstand conventional chemotherapeutic agents [3, 7]. We and others have been able to isolate GSCs with the stem cell-like properties of self-renewal, neurosphere formation, and ability of GSCs to recapitulate the genotype and phenotype of the tumor found in patients in immunocompromised mice [2, 3, 8–10]. The CSC marker CD133 (prominin-1) [3], once considered the discernible feature of GSCs, is now not so clear-cut as CD133-negative GSCs have also been shown to have stem cell-like properties [11, 12]. Although the association of CD133 with GSCs has also linked it to the radioresistance of GBMs, the drug resistance of GSCs, and the tumorigenicity of GSCs, very little is known regarding its function. Interestingly, Barcelos and colleagues demonstrated that CD133 progenitors were involved in healing with the activation of angiogenesis in diabetic ischemic ulcer patients, suggesting a role for CD133 in tumor angiogenesis [13]. Nevertheless it is clear that CD133 is associated with the phenotypic characteristics of GSCs that help them combat treatment. Having said that, although CD133 is an important GSC marker that can form tumors in immunocompromised mice, studies have also shown that other stem cell markers have been identified that confer at least some of the major characteristics found in CD133-expressing GSCs; such markers include CD15/SSEA1 [14], CD44 [15], and the ABCG2 transporter [7].

The Potential for Immunotherapy for Gliomas

The ultimate goal for any treatment regarding brain cancer would be eradication of the tumor; however a milestone that is believed to be achievable for targeting GSCs using immunotherapy is that it could eliminate GBM recurrence after treatment with conventional radiation and chemotherapy. Although a definitive stem cell marker has not been found in GSCs, the markers we do have certainly give us populations of GSCs that are involved in the progression and relapse of the disease. It is likely that even among these populations GSCs express distinct antigenicity providing opportunities for treatment by immunotherapy. This is evidenced by the fact that several proteins expressed in stem cells and that are also believed to be involved in GSCs are only seen in the early stages of development and glioblastomas; particularly GBM affects a population where targeting these developmental stage markers may be beneficial.

Treatments

There is a wide variety of compelling immunotherapeutic strategies that are currently being researched. These strategies are primarily categorized as passive immunotherapy, adoptive immunotherapy, and active immunotherapy. Passive immunotherapy takes advantage of immune effector molecules such as antibodies or toxins to target tumors without directly activating the immune system whereas adoptive immunotherapy, such as adoptive T cell transfer or use of T cells with chimeric antigen receptors (CARs), utilizes immune cells that have been stimulated ex vivo to react against tumor antigens and then re-administered to the patient in the hopes of a therapeutic benefit. Active immunotherapy, often referred to as a vaccination, involves stimulating the patient's immune system directly, using various sources of tumor-associated antigens (TAAs) such as tumor lysates, whole tumor cells, mRNAs, as well as natural and synthetic peptides. These antigens may or may not be coupled to dendritic cells.

Passive Immunotherapy

This therapy consists of therapeutic modalities that utilize immune-effector cells and/or a variety of molecules, including monoclonal antibodies (mAb) and cytokines to initiate an immune response. The most widely used form of this therapy involves the use of mAbs.

Antibody Therapy

In this method antibodies may be used naked, or as a platform to deliver toxic molecules, and are specific to the tumor antigens they target, binding with extremely high affinity [16]. However, the use of mAb presents a number of challenges when used against brain tumors. The size of mAbs favors delivery directly to the tumor as the molecules that pass across the BBB are preferably small and uncharged. While chemotherapy agents have a molecular weight of 1 kDa and rely on diffusion, IgG antibodies are 150 kDa and must rely upon convection, flowing down a pressure gradient, to be transported through the tissue [17–19]. As the interstitial pressure within a tumor has been shown to be much greater than surrounding tissue, the transport of antibodies may prove extremely difficult [18–20]. In addition the free antibody must be able to bypass target-free normal tissue, and not only bind to the tumor boundary layer but also diffuse into the tumor core. As the boundary contains a high concentration of the antibody, to reach the tumor core the antibody will also have to overcome this significant concentration gradient as well as the increased cell density in tumors. If these are not overcome the interior of the tumor will remain ineffectively treated.

For the most effective therapy, it is believed that the glioma cells being treated should have a low turnover time and exhibit an epitope with a minimum of 10^5 surface markers per cell for effective targeting, and that the antigen being targeted is glioma specific in an effort to preserve normal brain parenchyma [21]. In addition, to overcome delivery issues of antibodies through the tumor bulk, it is common that antibody therapy is accompanied by resection. While conventional delivery techniques include intratumoral injections and intraventricular infusions that rely on diffusion, a novel delivery technique that has emerged is convection-enhanced delivery (CED), which is able to enhance the delivery of both small and large molecules at a high pressure via an intracranial catheter [22]. While not the "magic bullet" in the fight against cancer, as suggested by Paul Ehrlich in 1906, targeted antibody therapy is still a promising immunotherapy technique.

Unlabeled Antibody Therapy

With the advent of tissue microarray analysis, serial analysis of gene expression (SAGE), and other techniques, identification of TAAs specific to gliomas has been achieved. These TAAs are targets of tumor-reactive T cells.

This has allowed mAbs to be raised that are able to bind with high specificity and affinity to antigens and initiate a biological response. This response may occur by activating antibody-dependent cellular cytotoxicity (ADCC), inducing C1 complement binding, or altering the signaling pathways that are necessary for tumor growth or apoptosis [23]. However, this response is influenced by a number of things including how accessible and stable the antigen is in the tumor. Antigen density is another factor that must be considered with this treatment as low levels of antigens would result in less targeting.

One form of tumor antigen is that of growth factors and their receptors. The overexpression of growth factors and their receptors on several types of tumors have been well studied and are known to play a significant role in tumorigenesis. Inhibiting these growth factors by blocking their receptors should, in theory, result in tumor regression. Epidermal growth factor receptor (EGFR) is overexpressed in over 50 % of high-grade gliomas making it an attractive therapeutic target [24]. EGFR is known to promote survival, tumor proliferation, as well as motility by activating downstream signaling pathways such as PI3K/Akt, RAS/MEK, and PKC [25]. Indeed, binding of EGFR to inhibit its activity was shown to inhibit tumor growth and induce apoptosis in GBM cell lines that overexpressed the receptor [26].

It was in 1996 that EMD55900 (an anti-EGFR mAb) was found to produce substantial nontoxic in vivo tumor binding in patients with malignant glioma [27].

Cetuximab (Erbitux), an anti-EGFR mAb that binds to the extracellular domain of EGFR, has been shown to be a fairly effective therapy. While targeting EGFR with mAbs has been shown to be effective in other cancers, i.e., trastuzumab (Herceptin), in breast cancer, and panitumumab (Vectibix) in colon cancer, their effectiveness in gliomas has not been well studied. But a new target may hold promise: the mutant variation of EGFR, EGFRvIII.

Given the genetic rearrangements that often occur within tumor cells, it is not surprising to find that this is the case with gliomas; EGFRvIII is the most common mutant. This mutant is an in-frame splice variant that contains a deletion from exons 2–7 resulting in a constitutively active tyrosine kinase which causes uninhibited proliferation, invasion, and inhibition of apoptosis [28]. In EGFRvIII the insertion of a novel glycine results in the fusion of normally distant parts of the epitope creating a tumor-specific epitope. EGFRvIII is restricted to cancer cells with more than 50 % of GBMs expressing this mutant variant, making it a novel target. There have been promising studies showing decreased tumor volume and increased survival in animal models of brain tumors that have been treated with anti-EGFRvIII mAbs [29–31].

Another growth factor that has shown great therapeutic potential is vascular endothelial growth factor (VEGF) which is significantly involved in cell survival, proliferation, migration, and invasion. VEGF plays a critical role in controlling the vascular permeability of the blood–brain barrier (BBB) as it is able to increase permeability. GBMs have been shown to have higher levels of VEGF in comparison to other malignancies, with high expression correlating to a poorer prognosis [32, 33]. Recently however, the highly publicized VEGFR inhibitor bevacizumab failed in phase III clinical trials in brain cancer [34]. This has led to a reevaluation of this drug and the concept of VEGF inhibition.

Radiolabeled Antibody Therapy

While unlabeled antibodies may confer some degree of therapy, their impact can be greatly enhanced when conjugated to a variety of effectors that may be delivered directly to the tumor cell. Radiolabeled nucleotides (radionucleotides) are the most frequent conjugate used in mAb therapy. In a manner similar to radiotherapy, radiolabeled antibodies deliver a lethal dose of radiation to the tumor; however the antibodies have the advantage of greater target specificity. Cytotoxicity occurs due to irreparable DNA damage that results from the ionization of emitted particles from a disintegration reaction [17]. In addition, studies as early as 1965 have shown that radionucleotides demonstrate the potential for enhanced diagnostic imaging of gliomas [35]. This form of therapy involving radionucleotide-conjugated mAbs is known as radioimmunotherapy (RIT). As RIT is still evolving a variety of radionucleotides are being investigated including but not limited to ^{225}Ac, ^{213}Bi, ^{125}I, ^{224}Ra, ^{188}Re, and ^{212}Pb. Improved targeting strategies and host response predictions are also being made as well as improvements in dosimetry. These improvements, however, should not overshadow the difficulties faced with radiotherapy. The short half-life of the radioisotopes necessitates the simple and speedy generation of RITs. Dosimetry (radiation absorbed by the body) itself is challenging due to differences in the clearance rates of the radionucleotides by patients as well as differences in the tumor geometry and characteristics.

The most widely studied tumor-associated antigen involved in RIT would be tenascin. An extracellular matrix glycoprotein, tenascin is prominently expressed on high-grade gliomas. This expression has been correlated to disease progression with over 95 % of GBMs displaying high levels of this protein [36]. This has made it a favorable target for glioma studies using RIT. Multiple antibodies against tenascin have been developed including 81C6, which was developed over 30 years ago by the Bigner group at Duke University as well as BC-2 which has been used in European clinical trials [37, 38].

Antibodies against EGFR and VEGF have also been conjugated to radionucleotides. Rosenkranz and colleagues developed a ^{211}At-conjugated internalizable ligand to EGFR which resulted in a more than tenfold increase in cellular toxicity of various cancer cell lines [39]. Other targets have included the extra domain of fibronectin, which is also a marker of tumor angiogenesis.

The use of antibodies has the advantages of being precise and specific, but there are substantial challenges in delivery and penetrance that remain in addition to the lack of a prolonged antitumor response.

Cytokine Therapy

Cytokine immunotherapy is a form of passive immunotherapy that can be further classified into a nonspecific immunotherapy. Unlike monoclonal antibodies which are able to elicit a specific immune response, cytokines are able to generate a wide range of responses with the rationale being that they upregulate antitumor effector functions especially those associated with TA-specific T cells. However this response is transient as in normal cytokine function the response is not meant to be prolonged and uncontrolled as this can result in autoimmunity and a wealth of other problems [40]. In truth cytokines are able to generate different effects in tumor target cells and effector lymphocytes. Some cytokines, including the interferons (IFNs), promote immunity and tumor destruction, whereas others, such as tumor growth factor beta (TGF)-β, inhibit immune function and promote tumor growth.

A variety of therapies with cytokines have been given, such as therapy against IL-12, IL-2, and IFNs. It should be noted that these therapies are often given alongside other immunotherapies such as IL-2 being used in adoptive therapy (discussed later). IL-2 is what generated the initial interest in cytokine therapy. It is able to bind to the surface of T cells to stimulate a response that leads to T cell proliferation, activation of other immune cells such as NK cells, and production of additional cytokines. Local delivery of IL-12 is also able to induce an immune response.

There have been a variety of cytokine receptors that have been found to be overexpressed in gliomas including IL-4 and IL-13Rα2. The use of cytokines to alter receptor–ligand interactions, and negatively affect necessary signaling pathways for the tumor, allows tumor cells to be targeted. In addition cytotoxicity may be achieved if the cytokines are fused to a toxin such as *Pseudomonas exotoxin* (PE). For example, IL-4 is overexpressed in glioma cells in relation to normal brain tissue. A chimeric fusion protein containing domains of IL-4 and PE was developed by Puri et al. that was found to be highly cytotoxic to IL-4R-bearing glioblastoma cells [41]. A composite of TGFα-PE (EGFR-binding ligand and PE cytotoxin) named TP38 has had quite promising results [42]. In addition cells may be genetically manipulated to express these cytokines in vivo to initiate an immune response. Cells have been transduced to express cytokines such as IL-1, IL-2, IL-3, IL-4, IL-6, IL-12, IFNγ-, TNF-α, granulocyte colony-stimulating factor, leukemia inhibitory factor, macrophage migration inhibitory factor, or GM-CSF. While tumor cells were initially manipulated to express these cytokines, the feasibility of such an act is not always possible as some patients do not have viable tumor cells. Therefore some researchers have engineered allogeneic cells such as fibroblasts to release these cytokines in vivo.

A recent study used the local delivery of IL-12 and IL-23 (a member of the IL-12 family) to induce an antitumor response to treat GBM in a syngeneic mouse model. Using a GL-261 mouse model with glioma cells engineered to express IL-12 and IL-23, they found that IL-12 but not 23 led to tumor clearance; this clearance was T cell dependent and elicited potent immunological memory [43]. To determine clinical relevance, the GBM was allowed to progress until life expectancy was less than 3 weeks before initiating therapy. IL-12 treatment alone led to tumor removal in only 25 % of the mice; however when this treatment was combined with blockade of CTLA-4 found on T cells, it caused a severe decrease in Treg and an increase in effector T cells. This resulted in tumor eradication in over 80 % of mice. The finding of new cytokines as well as novel combination strategies of cytokine delivery, such as that seen above, will be necessary for future directions of cytokine therapy in gliomas.

Adoptive Therapy

Adoptive immunotherapy is the process of ex vivo activation and expansion of effector cells obtained from the peripheral blood of the host patient, followed by re-administering them back into that patient in order to induce targeted destruction of actively proliferating tumor cells. The administration may be done intravenously or directly into the tumor. An old therapy utilized in the treatment of central nervous system (CNS) malignancies began with the simple reinfusion of a patient's peripheral blood mononuclear cells (PBMCs) into the resected area of their tumor. However, over time, treatment has shifted to the use of lymphokine-activated killer (LAK) cells and cytotoxic T lymphocytes (CTLs).

LAK Cells

Nonspecific effector cells that are a subpopulation of PBMCs are obtained from patients and activated ex vivo with the use of high concentrations of IL-2, a T cell growth factor, to become LAK cells that are able to induce antitumor effects [44, 45]. The LAK cells are not primed before reintroduction to target any specific tumor antigens but are dependent on host APC presentation to initiate a specified immune response against a tumor. In vitro studies have shown that these LAK cells are able to selectively lyse autologous tumor cells (ATCs) as well as allogeneic tumors with little cytotoxicity to normal brain tissue [46–48]. As LAK cells are not capable of migrating to the tumor location, it is necessary that they be injected at the area of surgical resection. This is often given with a low dose of IL-2 administered locally as given intravenously it results in very high toxicity. Disregarding the toxicities associated with LAK/IL-2 treatment, the primary problem causing low efficacy and minimal positive benefit in LAK cell therapy is likely the lack of tumor specificity. However, an in-depth understanding of exactly how antigen recognition occurs remains to be defined.

Cytotoxic T Lymphocytes

As discussed, LAK cells are cultured ex vivo and reintroduced to the patient alongside IL-2 but still rely on host APC presentation of tumor antigens for immune targeting. Since the tumor can evade host immune detection through a number of mechanisms, this presents an obvious shortcoming when using LAK cells for immunotherapy. CTLs are extracted from PBMCs and activated ex vivo against glioma antigens. In one method of activation, isolated PBMCs are cultured with host ATCs in order to prime them with tumor antigen before reintroducing them back into the patient. Alternatively, the patient may be vaccinated with ATCs in a process known as in vivo activation. The CTLs are then isolated post-vaccination and expanded ex vivo before being reintroduced to the patient. CTLs have also been derived from drainage of lymph nodes post-vaccination or by isolating tumor-infiltrating lymphocytes. In both cases, isolated cells were expanded ex vivo and reintroduced to the patient since some form of in vivo activation had already taken place. Of course the natural progression today is to consider genetically engineered options for adoptive T cell therapy in order to create the most specifically targeted immunotherapeutic approach. Artificially constructed antigen receptors have been engineered to target specific glioma peptides such as the previously discussed IL-13Rα2, HER2, and EGFRvIII [49–51]. These chimeric antigen receptors (CARs) allow researchers to manipulate the specificity of the injectable immune elements such as CTLs in a controlled manner and the high efficacy and power of the eliciting response has been demonstrated in animal models. A major issue with T cell therapies is damage to normal tissues as well as on-target, off-target toxicities given that tissues other than that of the tumor may express the targeted antigen. This damage can only be controlled by nonspecific immunosuppression or T cell elimination, both of which may eliminate any therapeutic benefit of the T cell therapy. To combat this, Fedorov et al. have recently engineered antigen-specific inhibitory chimeric antigen receptors (iCARs) that are able to limit cytokine secretion, cytotoxicity, and proliferation resulting in an initial constraint of the T cell response to activation by an endogenous T cell receptor or an activating CAR [52]. These results were found using CTLA-4- and PD-1-based iCARs. The inhibition is temporary, allowing T cells that have been switched off in the wrong tissues to function upon a subsequent encounter with the antigen recognized by their activating receptor. These self-regulating iCARs allow for a more specific targeting of tumor tissue versus healthy tissue.

Active Immunotherapy

More recent measures of treating gliomas have focused on active immunotherapy, a method relying on vaccines to stimulate the host's immune system. The ultimate goal of the vaccine is to elicit a response from one arm of the immune system such as antibodies or lymphocytes against the TAAs presented by the tumors. While there are multiple benefits to active immunotherapy, of importance is that the immune response be able to persist for a long time and that T cells can retain the memory that prevents the cancer from recurring.

Autologous Tumor Cell Vaccines

One approach to stimulating a vaccine effect is the use of ATCs that have been modified in such a way that they are unable to form tumors and also have increased immunogenicity. Whole cells, parts of cells such as the membrane, or simply antigens may be used that allow targeting of a specific tumor. ATCs used in vaccination are able to present a whole host of tumor cell markers to the immune system of the patient and in theory elicit a robust native immune response against tumors specifically. Patient-derived ATCs are attenuated before being reintroduced into the patient and are usually supplemented with other tumor-associated cytokines and signal molecules to maximize the stimulatory effect of the vaccine. Viral adjuvants may also be used as the Newcastle disease virus was recently used as an adjuvant for an ATC vaccine. The earliest active immunotherapy vaccines utilized irradiated ATCs that had either been engineered to secrete cytokines or combined with cytokines themselves. However this method was limited by a variety of factors including the poor antigen-presenting capacity of the cells as well as issues with quality control in regard to expansion of the tumor cells.

Dendritic Cell Vaccines

Using ATCs as antigen-presenting cells may work on some level but they are not the most efficient platform for vaccination. In CNS tumors, the glioma ATCs express relatively low levels of costimulatory molecules compared to "professional" APCs and therefore only elicit a dampened immune response. This has moved people towards a more promising strategy, using dendritic cell (DC)-based vaccines. Often termed "nature's adjuvant," DCs were identified by Steinman in the 1970s [53]. They originate in the bone marrow as hematopoietic stem cells and enter the blood as precursor cells where they ultimately seed tissue as nonproliferating immature DC cells. These immature cells are specialized in antigen detection and capture but are poor T cell activators. It is only after antigen capture that the DCs undergo a maturation process resulting in a potent ability to prime and activate T cells. Their ability to stimulate antigen-specific immunity, as well as their role as the most potent T cell activators (they are the only cells that can prime naïve T cells), makes DCs an essential factor in the immune response serving as a tie between the innate and adaptive immune responses.

DCs are natural targets for active immunotherapy as they come ready with MHC class I and class II molecules along with a host of costimulatory factors such as CD40, CD80, and CD86 [54]. In addition, the DC-expressed CCR7 receptor binds the CCL19 and CCL21 ligands which are highly expressed in the lymph nodes, facilitating antigen presentation at LNs to recruit a more robust immune response targeting tumor cells [55]. They are also able to produce a variety of cytokines including IL-12, which help send naive CD4 T cells towards a Th1 phenotype and type-1 IFN, which recruit more activated macrophages to allow phagocytosis. It was previously thought that these cells did not have a great role in the CNS as only rare isolated cells were reported across a variety of species but recent studies have shown extensive populations of MHC II expressing cells in the choroid plexus and other areas that appear to be DCs [56]. In addition T cells are able to cross the CNS in healthy animals [57]. DC-based vaccines therefore are powerful because DCs are great immune activators and are also capable of initiating tolerance [58].

One of the most widely studied use of DC's in regard to glioma is the use of DC vaccination. The DCs used are generated from the patient's peripheral monocytes and expanded ex vivo. Given the limitation of monocytes available, some groups have developed methods to generate DCs from pluripotent stem cells [59]. While earlier immunotherapy trials used immature DCs, the majority of DCs used in trials are pushed towards maturation using a variety of factors. Once the autologous DCs have expanded, they are loaded with TAAs from the tumor being treated. The TAAs, which are targets of tumor-reactive T cells, include tumor extracts, peptides, tumor DNA or RNA, and tumor cells fused using a tension-active compound such as polyethylene glycol. As previously mentioned some of the tumor-associated peptides include HER2 (ERB-B2), MAGE-1, IL-13Rα2, SOX2, EGFRvIII, EphA2, as well as SART1 and Survivin [60].

Primed with antigens, the DCs are then reintroduced into the patient and present the tumor antigens to boost the patient's immune system and elicit the appropriate

T cell response that will launch an immune attack on specific tumor cells. In addition, it may also result in a memory immune response, whereby it safeguards the body from tumor reoccurrence, a major obstacle faced in current therapies.

Peptide-Based Vaccines

One of the drawbacks of using whole cell-derived antigens such as those from tumor lysates is the possibility of autoimmune encephalitis occurring. In addition the difficulties faced in preparation as well as quality control issues give rise to valid limitations regarding this approach. An alternative and attractive form of therapy is peptide-based vaccines, which are fairly easy to manufacture and administer in comparison. These peptides are based specifically on the CTL epitopes of TAAs and may or may not be administered alongside autologous dendritic cells. Synthetic peptides may also be manufactured that code for TAA-derived CTL epitopes allowing for increased safety and feasibility in clinical trials as these peptides are less likely to induce autoimmunity and also have an "off-the-shelf" factor that is extremely useful. These peptides may be administered with or without dendritic cells (as mentioned above).

Peptides utilized in cancer vaccines are often composed of nine amino acids that are able to bind to an MHC class I antigen and elicit a CTL response against tumor cells. To form a vaccine the peptides are mixed with an adjuvant and subcutaneously administered every 7–14 days. While the exact mechanism of action is not known it is believed that the peptide is captured by APCs which after moving through lymph nodes present the peptide to circulating CTLs. As the CTLs have T cell receptors that correspond to the injected peptide they are able to recognize the APCs, become activated, and undergo clonal expansion. These activated CTLs are then able to migrate to the brain and eliminate glioma cells via recognition of the corresponding peptide on the tumor cell [60].

A wide range of peptides have been evaluated and their immunogenic potential has been shown in a variety of studies, with treatment resulting in an immune response that targets tumors. One such peptide is the tumor-specific mutated section of EGFRvIII, PEP3. In one study, when PEP3 was conjugated to keyhole limpet hemocyanin (KLH) there was an absence of tumor development in nearly 70 % of the mice [61]. Another peptide that shows promise is the glycoprotein IL-13Rα2 that is not expressed in the normal brain but is overexpressed by more than 80 % of malignant gliomas [62]. An HLA-A24-restricted CTL epitope was recently identified in this protein which will allow for a greater population that can be targeted [63]. The WT1 gene, originally identified as the cause of Wilms tumor, is an oncogene that is overexpressed in many solid tumors including gliomas. Involved in cell proliferation and apoptosis, this peptide is a novel therapeutic target and two peptides derived from WT1 ($WT1_{126-134}$ and $WT1_{187-195}$) were shown to stimulate a CTL response [64]. Cytomegalovirus (CMV) is another potential target as it has been found in both high-grade and low-grade gliomas [65]. Other proteins being studied include Survivin (an apoptosis inhibitor protein), HER2, SOX family of transcription factors, and EphA2, a tyrosine kinase receptor involved in oncogenesis.

A unique ability of tumors is their ability to undergo immunoediting resulting in the lowering or complete lack of expression of the TAAs being targeted. In addition, the antigenicity of individual tumor antigens may be extremely weak. In an effort to overcome this, many investigators are choosing to incorporate multiple peptides generated from TAAs into one vaccine creating a peptide cocktail. One example would be the combination of ephrinA2, YKL-40, gp100, and IL-13 receptor-α2 by Okada et al. which proved to be safe [66].

Immunotherapy Treatment of Cancer Stem Cells

Given the widely suspected role of CSCs in tumor initiation and propagation as well as their resistance to conventional treatments, treating them using immunotherapy may be an effective method of targeting this population. As favor towards these therapies emerges, and our knowledge of CSCs increases, there has been a rise in the number of studies testing this theory. However, exactly how well these stem cells will respond to immunotherapies, particularly glioma stem cells (GSCs), remains to be elucidated.

Recent data suggests that antibody targeting of GSCs may be an effective method. A recent study by Huang et al. targeted CD133 high CSCs in pancreatic and hepatic cancers in nude mice with an anti-CD133 antibody (which included anti-CD3 bound to cytokine-induced killer (CIK) cells) [67]. The addition of the anti-CD133 was able to significantly inhibit tumor growth when compared to CIK or CD3-CIK treatment alone. However, it should be noted that the combination of CD133-CD3-CIK was also a better inhibitor than merely using CD133-CD3 alone, giving further credence to the combination of immunotherapeutic treatments, which in this case was antibodies bound to effector cells.

In multiple myeloma, a bispecific antibody (C3B3) was used against a CSC like side population (SP) cells that exhibited resistance to chemotherapy. This antibody was able to induce cytotoxicity in this stem-like subpopulation of multiple myeloma cells, as well as suppress colony formation and inhibit tumorigenesis both in vitro and in vivo [68]. These results indicate the possible therapeutic benefit of using engineered antibodies against CSCs.

Indeed Schlaak et al. were the first to substantiate antibody targeting of CSCs in a single chemotherapy-refractory metastatic melanoma patient. Melanoma CSCs are characterized by chemotherapy resistance and CD20 overexpression; using the anti-CD20 antibody, Rituximab, they were able to achieve lasting remission of the melanoma, a decrease in the melanoma serum marker S-100 to physiological levels with very little toxicity [69].

Another study looked at CTL-mediated therapy using SOX6, a glioma antigen that has been found to be overexpressed in GSCs as well as gliomas. Vaccination with SOX6 DNA was capable of inducing a glioma-specific CTL response in a mouse model; furthermore SOX6-derived peptides (HLA-A2 and A24) were able to prime CTLs that were able to lyse GSC cells while normal cells were not affected [70].

The use of DC vaccines to target GSCs has also been indicated as a favorable therapy. As previously discussed, DCs may be primed with a variety of antigens including DNA or RNA. One study using glioma 9L tumorspheres (a CSC brain tumor model that includes expression of nestin and CD133) performed DC vaccination with 9L neurosphere lysate, but not conventionally cultured 9L cells, to induce a CTL response, which recognized GSCs and increased the survival of the rats bearing 9L gliomas [71]. In a similar study, DCs transfected with RNA from these spheres were able to induce a strong antitumor response and significantly inhibit glioma growth and also prolonged the survival of glioma-bearing rats [72]. While there are advantages to using RNA including the need for only a small amount of tissue as RNA can be amplified from very few cells as well as the fact that the identity of GSC-specific tumor antigens does not need to be known, there are also drawbacks. One significant downside is the fact that the RNA is easily degraded.

Another study highlighting the benefits of using GSCs was able to show that pulsing DCs with GL261 neurospheres (NS) was more efficient and had a stronger antitumor effect than pulsing with GL261 adherent tumor cells (AC). DCs loaded with GL261-NS (DC-NS) were able to cure 80 % and 60 % of GL261-AC and GL261-NS tumors, respectively, whereas DC-AC cured only 50 % of GL261-AC tumors and none of the GL261-NS tumors [73]. Despite there being no characterization of the tumor antigens, it does put forth the potential that lies within GSCs.

While many CSC studies use immunosuppressed mice, Ning et al. took a different approach and used syngeneic immunocompetent mice to examine the vaccination effects produced in these mice by CSC populations from genetically distinct tumors. Immunocompetent mice were used as the authors believe that immunosuppressed mice prevent an assessment of the immunologic interactions and effects of CSCs [74]. They found that priming DCs with tumor cells enriched for CSCs conferred antitumor immunity by inducing both humoral and cellular responses that directly targeted CSCs through complement-dependent cytotoxicity and CTLs. In addition the CTLs obtained from the vaccinated host were also capable of killing CSCs in vitro.

The majority of cancer immunotherapies have targeted differentiated tumor cells. However, it must be taken into consideration that these antigens may be selectively expressed on differentiated cells and not on CSCs. Thus in addition to targeting known tumor antigens work is being done to reveal novel antigens that are specific to CSCs. In a study looking at cancer testis antigen (CTA) genes in GSCs isolated from glioma cell lines and tissues it was found that the CTA genes were highly expressed in GSCs in comparison to differentiated tumor cells. The expression of one gene, LAGE-1, was limited to the CSCs which may indicate a candidate for a CSC-specific antigen [75, 76]. Elucidating CSC-specific antigens will help prevent CSCs that are capable of escaping immunological interventions.

Based on the results of these studies using CSCs and immunotherapy, it appears that as we clarify the differences between CSCs, normal stem cells, and non-stem cancer cells, we will be better able to target CSCs for both mono- and combination therapy.

Future Treatment Directions

The field of immunotherapy to treat gliomas is rapidly evolving; new techniques and strategies are steadily emerging as the concept takes root. Although the strategies mentioned are fairly young and hold significant promise their effect to date has only been moderate and has yet to be successful among the masses. This is in part due to the fact that with new therapies, there are drawbacks that must be overcome including issues with the treatments themselves as well as tumor heterogeneity: additionally the presentation of tumor antigens remaining at an unsatisfying level of efficiency as well as immune evasion by the tumor via induction of regulatory T cells (Tregs) and myeloid-derived suppressor cells (MDSCs). Thus additional strategies should include efforts that deplete Tregs or MDSCs allowing for an enhanced immune response. Another major issue facing immunotherapy in the CNS is the possibility of developing an autoimmune response. While the ideal antigen would be tumor specific the majority of antigens targeted are tumor associated; thus they are present in varying levels on normal tissues increasing the possibility of autoimmunity. Another interesting area to target would be the activation of the immune cells inherent to the CNS to determine if they could indeed elicit an immune response as well as how effective this response would be. Identification of biomarkers that may determine patients who will gain the most benefit from immunotherapy strategies is another promising area of study. It remains to be determined if the benefits of a combination treatment will be time or age dependent; should it be given to patients who have advanced-stage glioma or will it only be effective in the initial stages? Younger versus older patients? Identification of these prognostic indicators can help to design more specific clinical trials. However, given that GSCs are believed to reinitiate tumors, targeting this subpopulation may be the ultimate key to successful immunotherapy.

Although the immunotherapies discussed, as well as those that were not, may not be effective as a single-modality treatment, we believe that the most effective immunotherapies will be a cocktail that combines three or more different immuno/chemotherapies alongside TMZ. In essence, by combining synergistic treatments, immune evasion will be decreased allowing for a CTL-mediated antigen-specific tumor elimination. Transfecting autologous dendritic cells with autologous GSC-mRNA for example, in addition to cytokine therapy, may prove effective and prolong recurrence-free survival. The current results seen in clinical trials strongly support the use of immunotherapy as an adjuvant treatment alongside standard glioma care. Additionally the use of histone deacetylase inhibitors alongside TMZ and immunotherapy may enhance survival as studies have shown that these inhibitors significantly improved survival in TMZ-treated patients. It is the hope that one day the primary benefit of immunotherapy is that in addition to an initial immune response, the immune system may be able to adapt to the oft-occurring changes in the tumor cells and microenvironment, bypassing any resistance that usually occurs with drugs.

Approaches utilizing immunotherapy to managing cancer treatment are the result of finding ways to exploit the body's own immune system, through passive immunotherapy, adoptive immunotherapy, or active immunotherapy. There has been a concerted effort to harness these treatment modalities for clinical use. The rationale for pursuing such ends in the clinic has been slowly established in the lab for several decades. The tumor microenvironment has evolved in such a way that allows for cancers today to effectively evade host immunosurveillance and at times redirect the host response into self-preservative functions, utilizing a wide array of tactics including antigen mutations, responsive downregulation of certain target antigens, and even selective survival of certain robust tumor subpopulations. GBM is particularly relevant in this discussion since it represents one of the most heterogeneous tumor subclasses making them incredibly difficult to clinically target.

Clinical Trials

Immunotherapy Clinical Trials in High-Grade Gliomas: Historical

Immunotherapy remains a very desirable method of attacking complicated and advanced tumors. Over the past 25 years a number of Phase I and Phase II clinical trials for high-grade gliomas have been conducted using a variety of approaches with varied results, including adoptive T cell therapies, ATC and DC vaccines, and more recently vaccines with tumor peptide adjuvants.

The first report of LAK-mediated immunotherapy in CNS patients came in 1986 from Jacobs et al. where six high-grade glioma patients undergoing a tumor-debulking craniotomy were directly injected with LAK cells at the tumor site [77]. Although the treatment was well tolerated, there were no clinically significant responses versus normal disease progression.

The primary complication with LAK cell injection was cerebral edema, reported in several trials where there was intracavitary or intralesional delivery of LAK + IL2 [78–84]. In two separate studies conducted in 1988, Merchant et al. reported increased intracranial pressure inducing cerebral edema in nearly every subject. ICP was always elevated even when compared to the expected levels of edema expected post-craniotomy. Barba et al. showed similar results in 1989, additionally reporting signs of neurological deficits resulting from cerebral edema in over half of the patients post-intracavitary injection of LAK + IL-2. Both studies determined that the safety of injecting LAK cells with IL-2 should be carefully evaluated before proceeding, minding not to cause excessive swelling in areas of the brain more sensitive to changes in pressure (e.g., thalamus and brainstem) [78]. It is also difficult to access the relative success of LAK-mediated immunotherapy due to the variance in reported results. Early trials such as Merchant et al. reported mean survival at around 25–30 weeks from the time of pre-immunotherapy surgery with no benefit

[81, 82]. Of the 33 patients enrolled between two studies in 1988, 30 were either GBM or recurrent GBM. In 1988 Yoshida et al. also conducted a Phase I LAK + IL2 trial but in 23 patients with recurrent anaplastic astrocytoma (AA) [85]. Of the 23 patients, 6 were reported to have responded to immunotherapy. Also in 1988, Okamoto et al. conducted a pilot study with six young patients with medulloblastoma with CSF dissemination with positive results [86]. Three patients showed improved neurological signs with negative detection of malignant cells in the CSF, with one patient demonstrating full response within 20 months posttreatment.

Hayes et al. (2001) reported 28 patients in a Phase I/II trial for recurrent high-grade gliomas, again using LAK + IL-2 via intracavitary injection [44]. One patient reported a complete response while three others had either a partial or minor response to the treatment. Median overall survival was reported at 53 weeks in GBM patients versus 26 weeks in non-treated controls, showing significant positive improvement but only in a small subset of the patient pool. Two of the patients that did respond remained alive and nonrecurrent after 8 years posttreatment at the time of the report. Dillman et al. (2004) also reported increased overall survival (17.5 months) in 31 recurrent GBM patients out of 40 enrolled and compared to 41 age-matched controls (13.6 months) [87]. They again reported another Phase II clinical trial in 2009, this time enrolling 33 newly diagnosed GBM patients as opposed to recurrent GBM patients. Expectedly, the newly diagnosed patients responded better to the same LAK treatment as compared to recurrent GBM patients, with a reported median overall survival of 20.5 months [88].

Although LAK cell-mediated immunotherapy has shown improved signs in more recent trials, the efficacy and positive benefit in advanced CNS tumors is not definitive. Moreover, the success of the treatment seems to be intricately related to the tumor type and perhaps also patient age. Primary complications include neurological irritation, fever, and cerebral edema as mentioned earlier. Many of these problems, notably the cases of edema, may well be tied to the direct injection of an inflammatory cytokine like IL-12 into CNS tissue. In any case, the majority of adoptive T cell therapy is no longer done with LAK cells today, but rather with the use of CTL.

In 1987 Kitahara et al. reported two out of five patients responding partially to intracavitary injection of CTLs with more than 50 % tumor regression in both cases [89].

Ten years later Kruse et al. conducted a Phase I trial with five patients of mixed recurrent tumors (GBM, A-ODG, AA). Of the five patients, three showed stable disease posttreatment but these did not include any GBM patients [90]. In this study IL-2 was also used in treatment as an adjuvant alongside the CTLs that were not derived from patient PBMC but rather sourced from PBMCs of MHC disparate donors primed with patient lymphocytes. Again patients that responded showed stabilized disease but no regression of the tumor itself. Tsurushima et al. (1999) showed three partial responses in both recurrent GBM and AA from a total of four enrolled patients while more recently Tsuboi et al. (2003) enrolled ten high-grade glioma patients and showed one complete response and three partial responses to treatment [91, 92]. Both of these cases showed relatively high tumor response rates (~50 %).

A 1999 trial by Quattrocchi et al. used tumor-infiltrating lymphocytes (TIL) with IL-2 adjuvant in three recurrent GBM and three recurrent AA patients. After a 12-month follow-up one AA patient showed a partial response and one showed complete response while only one GBM patient showed a complete response [93] all with only minor side effects.

Several trials have utilized CTLs derived from patients post-ATC vaccination [94–97]. In 2000 Sloan et al. conducted a pilot study with 16 recurrent GBM, 2 recurrent AA, and 1 recurrent gliosarcoma (GS) [97]. Of these patients one displayed complete response while seven had a partial response and nine more showed stable disease. Furthermore, 8 of the 19 patients had a radiological response and this response correlated positively with improved survival with a median surgery-free survival of 12 months posttreatment. Holladay et al. had opposite results 4 years earlier when a 15 patient pool of recurrent GBM and AA had no response throughout and progressive disease in all 15 patients [94]. In 1998 Plautz et al. showed three partial responses and four patients with survival over 12 months in a Phase I study with nine recurrent GBM and one recurrent AA [95].

Two years later in 2000, the same group was able to show four partial responses and two stable disease responses from a 12 patient pool of newly diagnosed GBMs and AAs [96].

There is clear variance in many of the results among all of the CTL-mediated therapies and some of that lies in the lack of standardization in the methodology itself. There has yet to be defined an optimal parameter for adoptive immunotherapy and the number of cells injected into patients in many clinical trials does in fact vary anywhere from 3×10^7 cells to 1×10^{11} cells. Kronick et al. recently developed a mathematical model to look at just this trying to account for the dynamic interaction between the immune elements such as lymphocytes and cytokine and the tumor microenvironment [99]. Based on this model the optimal system of immunotherapy in GBM would require at least 3×10^8 alloreactive CTLs on a 4-day injection regimen or 2×10^9 CTLs on a 5-day injection regimen depending on the tumor size. This would imply the primary reason for much of the failure to illicit a clinically significant response in many of the clinical trials so far was the dosage administered, which was on average 20-fold lower than required.

Recent studies have demonstrated the presence of cytomegalovirus (CMV) in human glioma tumors but not in the surrounding healthy tissue, suggesting CMV to be a therapeutic target that could be utilized in a CAR system [100, 101]. Furthermore, Dziurzynski et al. demonstrated that $CD133^+$ GSCs not only expressed CMV but also produced measurable quantities of CMV IL-10. This immunosuppresses human monocytes through multiple mechanisms, including upregulation of the immune-inhibitory molecule B7-H1 [102]. Taken together, all of these recent findings give good evidence to suggest promise in targeting CMV^+ tumor cells in the CNS to eradicate malignant progression and perhaps even destroy CSC propagation.

Apart from adoptive therapeutic techniques, there is ample work being done to suggest that active immunotherapy could provide the potential to additionally target tumors by stimulating the adaptive immune response. This is primarily accomplished by vaccinating patients against tumor antigens in order to elicit this response. Clinical trials to date have utilized ATCs, dendritic cells (DCs), and tumor peptides to vaccinate patients in this approach, each with its own benefits and each with varying success. An early 2004 pilot study by Ishikawa et al. used adjuvant tuberculin microparticles along with formalin-fixed ATC-based vaccine in 12 GBM patients. Of the patient pool five responded with a significant increase in median survival, up from 5 months to 20.3 months, while similar studies have shown at minimal establishment of stable disease [103]. Several trials have supplemented ATC-based vaccines with different adjuvants such as granulocyte-macrophage-stimulating factor (GM-CSF), Bacillus Calmette-Guérin, and Newcastle disease virus [94–98, 104, 105].

One of the first Phase I DC-based trials in CNS tumors was done by Kikuchi et al. in 2001 with seven patients using a tumor cell+DC fusion vaccine [106]. Of the five GBM patients, four maintained stable disease while one had a minor response. Two more of the remaining patients (AA and A-ODG) also maintained stable disease. Concurrently we (Yu et al.) did a Phase I trial with nine patients of newly diagnosed GBM and AA using acid-eluted MHC class I-associated tumor-specific peptides to prime DCs [107]. Results were positive with median overall survival increasing in the respondent group from 257 days to 455 days. In 2004 we repeated the trial but included an IL-12 adjuvant in a 15-patient Phase I trial. Four of the patients had a primary response of over 50 % tumor regression while two more maintained stable disease and one patient had a minor response. We (Yu et al.) also started a new Phase I trial that year with 14 patients with GBM or AA using DCs pulsed with tumor lysate [108].

Even in recurrent GBMs, median survival increased significantly from 30 weeks to 133 weeks. A 2008 Phase II trial by Wheeler et al. with 34 GBM patients again using DCs pulsed with tumor lysate showed three complete responses and one partial response to treatment with vaccine responder median survival up from 430 days to 642 days [109]. Furthermore, this study showed that patient survival was positively impacted by a treatment protocol that included both vaccination and chemotherapy, as opposed to any one of these treatments in isolation. All of these trials showed some level of positive benefit in the short term, especially when complemented with traditional treatment methods like surgical resection and chemotherapy. This was accomplished with minimal reported toxicities.

There is an explicit modulation of the immunological response detectable in the brain in patients who undergo DC-based vaccination. The measurement of such a modulation may require many different factors such as IFN-γ production, $CD4^+$ and $CD8^+$ precursor frequencies, and proportions of these precursors that also produce IFN-γ. These were the parameters for immunological function that were used in Fadul et al. post-DC vaccination using DCs primed with tumor lysate and injected into the bilateral cervical lymph nodes in ten newly diagnosed GBM patients [110].

Recent Immunotherapy Trials of GBM

Multiple Phase II clinical trials with newly diagnosed GBM patients were conducted in 2012. Cho et al.'s was a randomized, double-blind, placebo-controlled trial that enrolled 34 patients with 16 control subjects and 18 treatment group subjects that received DC vaccination primed with tumor lysate [112]. Median overall survival in the treatment group was 31.5 months, up from 15 months in the control. Jie et al.'s was an open-labeled, unrandomized trial that enrolled 25 patients with 12 controls and 15 treatment subjects receiving DC vaccine primed with heat-shocked tumor cells [111]. Once again median survival was improved in the treatment group at 17.5 months compared to the 10.5 months of the control. More studies are being conducted now into the possible safety and efficacy of using DC-based vaccine along with adjuvant chemotherapy, as many of the recent trials have suggested a possible synergism between the two treatments.

To date only one Phase II clinical trial for newly diagnosed GBM has provided evidence of improved survival outcome combining an experimental drug with standard radiation and chemotherapy. Dr. Manfred Westphal and colleagues from the Department of Neurosurgery at UKE Hamburg disclosed final results from a randomized Phase II trial of nimotuzumab for newly diagnosed GBM in addition to standard radiochemotherapy with temozolomide versus radiation and temozolomide alone at the 2012 Annual ASCO meeting [113]. The study showed a marked survival improvement from 15 months to 19.6 months in patients with an unmethylated MGMT promoter with 28 experimental and 28 control patients. ImmunoCellular Therapeutics presented preliminary data at the 2014 Annual ASCO meeting from their randomized Phase II trial of ICT-107 vaccine for newly diagnosed GBM patients [114]. All patients underwent standard radiochemotherapy and temozolomide following surgery and were positive for either human leukocyte antigen-A1 or -A2 (HLA-A1, HLA-A2). A total of 124 patients were enrolled and randomized to ICT-107 or placebo control. Progression-free survival in the vaccinated group of MGMT methylated, HLA-A2 positive was reported to be 24.1 months compared to 8.5 months in the placebo. These results represent the most compelling evidence of vaccine efficacy in affecting PFS of MGMT-methylated patients in any Phase II or Phase III trial to date.

There is a very valid risk of autoimmunity in any of these previously described trials due to the general overlap between targets within tumor lysate and normal tissue. Therefore it is more clinically useful to pursue even more targeted vaccines that spare damage to normal non-tumor tissue, especially in a delicate closed system such as the central nervous system. Recent work has focused on specific tumor peptides that are absent on normal tissue but still are effective targets. In 2005 Yajima et al. conducted a Phase I trial enrolling 25 patients either HLA-A24$^+$ or HLA-A2$^+$ with recurrent high-grade gliomas [115]. Each patient's PBMCs and blood plasma were tested against 23–25 peptides before being vaccinated with up to 4 peptides, positive for a detectable immune response along with Montanide ISA51 as an adjuvant.

Of the 25 patients, 5 showed partial response to vaccination and 9 more displayed stable disease. The overall survival of patients with recurrent GBM was 622 days. The post-vaccination tumor cavity of all positively responding patients showed significantly elevated levels of peptide-specific IgG.

Other tumor-specific peptide treatments have shown similarly promising results. Morita et al. conducted a Phase I/II trial in 2006 with 6 HLA-A2402$^+$ patients with Wilms tumor 1 (WT1)-positive tumors [117]. An HLA-A2402-restricted modified 9-mer WT1 peptide was used with a Montanide ISA51 adjuvant. One patient showed partial response and four more had stable disease. Izumoto et al. would repeat a similar study 2 years later in 2008 with a larger patient pool of 21 this time, all recurrent GBMs. Results were similar with two partial responses and ten stable diseases with a median survival of 36.7 weeks [116].

Sampson et al. would conduct two Phase II multicenter trials in 2010 and 2011 focusing this time on EGFRvIII as a target. Both studies were done with newly diagnosed GBM patients [118, 119]. In the 2010 study, 18 newly diagnosed GBMs were vaccinated with modified EGFRvIII peptide (PEPvIII KLH) and showed improved survival with a median overall survival of 26 months, versus 15 months in the control. In 14 patients tested, 6 showed increased humoral response against EGFRvIII and correlated to an even more prolonged median overall survival of 47.7 months. In 2011 the same study was conducted again in 22 patients with newly diagnosed GBM and similarly, medial overall survival increased to 23.6 months versus 15-month controls [121]. In 2013 Crane et al. used heat-shock protein peptide complex (HSPPC)-96 as a vaccination target in 12 current GBM patients [120]. Eleven of the 12 patients showed positive benefit and specific immune response to HSP-96-bound peptides. Median overall survival increased markedly to 47 weeks compared to 16 weeks and one patient who did not respond. A Phase I trial of tumor-associated antigen-pulsed DC vaccination for 22 enrolled patients with brain stem gliomas or GBMs was recently completed at Cedars Sinai Medical Center and the results are pending. More work is being actively pursued as these trials have demonstrated an improved efficacy with peptide-based vaccines.

In early 2013 Vik-Mo et al. published results of a DC vaccine study targeting CSCs in solid tumor GBMs [121]. In this trial monocyte-derived autologous DCs were transfected with mRNA from CSCs procured from each patient. Autologous GSC cultures were developed from 10 of 11 tumors from the initial trial enrollment. Of these patients, only seven were able to come off corticosteroids well enough to initiate immunotherapy. DC vaccination was introduced after standard 6 weeks of postoperative treatment of radiation and chemotherapy and good immune induction were observed in all patients. With no considerable adverse effects, progression-free survival improved nearly threefold, from 236 days in the control group to 694 days in the treatment group. This was a good example of a case where GBM vaccination successfully improved progression-free survival in a safe and well-tolerated manner, making the argument that therein lies great potential in using enhanced immune response mechanisms to attack resilient tumor subpopulations such as GSCs.

Antibodies themselves have been utilized with some success for some time in targeting gliomas. The feasibility of moving such cellular components across the blood–brain barrier has been established for some time after imaging work in glioma patients used radiolabeled antibodies to hone in on gliomas for image resolution [35]. The crossing of the blood–brain barrier is a very essential question to address when considering any type of cell-mediated therapy for CNS tumors and applies to all of the therapeutic approaches that have been discussed thus far.

Secondary to the deployment concern is the actual targeting design of the antibodies themselves. Tenascin is an extracellular matrix molecule featured prominently in high-grade gliomas. Tenascin-specific monoclonal antibodies (mAbs) have been developed both in the USA (81C6) and in Europe (BC-2) and have been adopted into clinical trials with varied success [122, 124]. Another more recent target has been EGFR in malignant gliomas, which has been shown to inhibit growth and induce tumor apoptosis when bound [26, 123]. A chimeric EGFR mAb called cetuximab was indicated to be effective in preclinical trials but failed to confer any significant clinical benefit in terms of progression-free survival or overall survival when it progressed to Phase II trials [125]. A small study has been done for other EGFR mAbs such as trastuzumab (Herceptin) and panitumumab (Vectibix) applied to high-grade gliomas despite well-established clinical benefits in certain forms of colon and breast cancer.

This however seems to be a more recent trend of unsuccessful translation of preclinical studies in CNS immunotherapy to the clinic. Bevacizumab (Avastin), a humanized monoclonal antibody (93 % human and 7 % murine sequences), that showed extremely promising preclinical results with evidence of prolonged overall survival in patients is a case in point. Several Phase I/II clinical trials published positive clinical outcomes in newly diagnosed GBM patients that underwent a course of bevacizumab plus temozolomide and radiotherapy post-debulking surgery [32, 126, 127]. Bevacizumab targeted VEGF activity in order to interfere with tumor proliferation by interfering with tumor angiogenesis. In particular, bevacizumab was able to undergo high-affinity binding to all isoforms of human VEGF, thereby counteracting VEGF's biological activity as it sterically blocks binding to the VEGFR-1 or VEGFR-2, thereby inhibiting any induction of signaling pathways [126]. First used in patients by Dr. Stark-Vance, this drug was approved in 2009 by the Food and Drug Administration (FDA) for recurrent GBMs. It is often used in combination with other therapies but further studies are necessary to delineate the mechanisms behind this synergy in treatment. Early Phase II trials seemed to suggest that when given alone or in conjunction with some chemotherapeutic, such as irinotecan, patient response rates and overall survival improved [127]. However, Phase III results from the most recent trial RTOG 0825 were reported at the meeting of the American Society of Clinical Oncology (ASCO) in late 2013 and suggested that although some progression-free survival was observed, most patients were seen to show negative response and decreased overall survival with an additional increase in treatment toxicity [129]. These results were soon corroborated by similar results from the AV Algio European GBM trial [128]. This promising antiangiogenic targeting mAb failed to deliver any significant clinical benefit to GBM patients and at times did more harm than good in terms of survival outcome.

The primary challenge when determining broad efficacy of CNS tumor immunotherapy in clinical trials is grounded in the great variance in trial design and targeting methodologies. In one regard it is expectedly hard to standardize treatment protocols for something as complicated as cancer immunotherapy in the brain since the heterogeneity of trial designs and treatment protocols simply reflects the intrinsic heterogeneity of the disease in question based on the knowledge we have today.

References

1. Allegra A, Alonci A, Penna G, Innao V, Gerace D, Rotondo F, Musolino C. The cancer stem cell hypothesis: a guide to potential molecular targets. Cancer Invest. 2014;32(9):470–95. doi:10.3109/07357907.2014.958231.
2. Bao S, Wu Q, McLendon RE, et al. Glioma stem cells promote radioresistance by preferential activation of the DNA damage response. Nature. 2006;444:756–60.
3. Singh SK, Hawkins C, Clarke ID, et al. Identification of human brain tumor initiating cells. Nature. 2004;432:396–401.
4. Wang R, Chadalavada K, Wilshire J, et al. Glioblastoma-stem like cells give rise to tumor epithelium. Nature. 2010;468:829–35.
5. Cheng L, Huang Z, Zhou W, et al. Glioblastoma stem cells generate vascular pericytes to support vessel function and tumor growth. Cell. 2013;153(1):139–52.
6. Lathia JD, et al. High-throughput flow cytometry screening reveals a role for junctional adhesion molecule as a cancer stem cell maintenance factor. Cell Rep. 2014;6(1):117–29.
7. Bleau AM, Hambardzumyan D, Ozawa T, et al. PTEN/PI3K/Akt pathway regulates the side population phenotype and ABCG2 activity in glioma tumor stem-like cells. Cell Stem Cell. 2009;4(3):226–35.
8. Ignatova TN, Kukekov VG, Laywell ED, et al. Human cortical glial tumors contain neural stem-like cells expressing astroglial and neuronal markers in vitro. Glia. 2002;39:193–206.
9. Xiangpeng Y, Curtin J, Xiong Y, et al. Isolation of cancer stem cells from adult glioblastoma multiforme. Oncogene. 2004;23:9392–400.
10. Lee J, Kotliarova S, Kotliarov Y, et al. Tumor stem cells derived from glioblastomas cultured in bFGF and EGF more closely mirror the phenotype and genotype of primary tumors than do serum-cultured cell lines. Cancer Cell. 2006;9:391–403.
11. Wang J, Sakariassen PO, Tsinkalovsky O, et al. CD133 negative glioma cells form tumors in nude rats and give rise to CD133 positive cells. Int J Cancer. 2008;122(4):761–8.
12. Joo KM, Kim SY, Song SY, et al. Clinical and biological implications of CD133-positive and CD133-negative cells in glioblastomas. Lab Invest. 2008;88(4):808–15.
13. Barcelos LS, Duplaa C, Krankel N, et al. Human CD133+ progenitor cells promote the healing of diabetic ischemic ulcers by paracrine stimulation of angiogenesis and activation of Wnt signaling. Circ Res. 2009;104(9):1095–102.
14. Jin Son M, Woolard K, Nam D-H, et al. SSEA-1 is an enrichment marker for tumor-initiating cells in human glioblastoma. Cell Stem Cell. 2009;4(5):440–52.
15. Anido J, Saez-Borderias A, Gonzalez-Junca A, et al. TGF-β receptor inhibitors target the $CD44^{high}/Id1^{high}$ glioma-initiating cell population in human glioblastoma. Cancer Cell. 2010;18:655–68.
16. Johnson LA, Sampson JH. Immunotherapy approaches for malignant glioma from 2007 to 2009. Curr Neurol Neurosci Rep. 2010;10(4):259–66.
17. DeVita VT, Hellstrom S, Rosenberg SA. Biologic therapy of cancer, vol. 2. Philadelphia: J. B. Lippincott Company; 1995.

18. Butler TP, Grantham FH, Gullino PM. Bulk transfer of fluid in the interstitial compartment of mammary tumors. Cancer Res. 1975;35(11 Pt 1):3084–8. PubMed: 1182701.
19. Jain RK, Baxter LT. Mechanisms of heterogeneous distribution of monoclonal antibodies and other macro-molecules in tumors: significance of elevated interstitial pressure. Cancer Res. 1988;48(24 Pt 1):7022–32. PubMed: 3191477.
20. Williams LE, Duda RB, Proffitt RT, Beatty BG, Beatty JD, Wong JY, Shively JE, Paxton RJ. Tumor uptake as a function of tumor mass: a mathematic model. J Nucl Med. 1988;29(1):103–9.
21. Wikstrand CJ, Cokgor I, Sampson JH, et al. Monoclonal antibody therapy of human gliomas: current status and future approaches. Cancer Metastasis Rev. 1999;18(4):451–64.
22. Bobo RH, Laske DW, Akbasak A, Morrison PF, Dedrick RL, Oldfield EH. Convection-enhanced delivery of macromolecules in the brain. Proc Natl Acad Sci U S A. 1994; 91(6):2076–80.
23. Cragg MS, French RR, Glennie MJ. Signaling antibodies in cancer therapy. Curr Opin Immunol. 1999;11(5):541–7.
24. Rivera F, Vega-Villegas ME, Lopez-Brea MF, Marquez R. Current situation of panitumumab, matuzumab, nimotuzumab and zalutumumab. Acta Oncol. 2008;47(1):9–19.
25. Ohno M, Natsume A, Ichiro Iwami K, et al. Retrovirally engineered T-cell-based immunotherapy targeting type III variant epidermal growth factor receptor, a glioma-associated antigen. Cancer Sci. 2010;101(12):2518–24.
26. Eller JL, Longo SL, Hicklin DJ, et al. Activity of antiepidermal growth factor receptor monoclonal antibody C225 against glioblastoma multiforme. Neurosurgery. 2002;51(4):1005–14.
27. Faillot T, et al. A phase I study of an anti-epidermal growth factor receptor monoclonal antibody for the treatment of malignant gliomas. Neurosurgery. 1996;39:478–83.
28. Wikstrand CJ, McLendon RE, Friedman AH, Bigner DD. Cell surface localization and density of the tumor associated variant of the epidermal growth factor receptor, EGFRvIII. Cancer Res. 1997;57(18):4130–40.
29. Zalutsky MR, Boskovitz A, Kuan CT, et al. Radioimmuno targeting of malignant glioma by monoclonal antibody D2C7 reactive against both wild-type and variant III mutant epidermal growth factor receptors. Nucl Med Biol. 2012;39(1):23–34.
30. Yang W, Barth RF, Wu G, et al. Development of a syngeneic rat brain tumor model expressing EGFRvIII and its use for molecular targeting studies with monoclonal antibody L8A4. Clin Cancer Res. 2005;11(1):341–50.
31. Perera RM, Narita Y, Furnari FB, et al. Treatment of human tumor xenografts with monoclonal antibody 806 in combination with a prototypical epidermal growth factor receptor-specific antibody generates enhanced antitumor activity. Clin Cancer Res. 2005;11(17):6390–9.
32. Vredenburgh JJ, Desjardins A, Reardon DA, et al. The addition of bevacizumab to standard radiation therapy and temozolomide followed by bevacizumab, temozolomide, and irinotecan for newly diagnosed glioblastoma. Clin Cancer Res. 2011;17(12):4119–24.
33. Mendelsohn J, Baselga J. The EGF receptor family as targets for cancer therapy. Oncogene. 2000;19(56):6550–65.
34. Gilbert MR, Dignam J, Won M, et al. RTOG 0825: phase III double-blind placebo-controlled trial evaluating Bevacizumab (Bev) in patients (Pts) with newly diagnosed glioblastoma (GBM). J Clin Oncol. 2013;31(suppl;abstr 1).
35. Day ED, Lassiter S, Woodhall B, Mahaley JL, Mahaley MS. The localization of radioantibodies in human brain tumors. I. Preliminary exploration. Cancer Res. 1965;25(6):773–8.
36. Zalutsky MR. Targeted radiotherapy of brain tumours. Br J Cancer. 2004;90(8):1469–73.
37. Bourdon MA, Wikstrand CJ, Furthmayr H, Matthews TJ, Bigner DD. Human glioma-mesenchymal extracellular matrix antigen defined by monoclonal antibody. Cancer Res. 1983;43(6):2796–805.
38. Riva P, Arista A, Franceschi G, et al. Local treatment of malignant gliomas by direct infusion of specific monoclonal antibodies labeled with 131I: comparison of the results obtained in recurrent and newly diagnosed tumors. Cancer Res. 1995;55(23 Suppl):5952s–6.

39. Rosenkranz AA, Vaidyanathan G, Pozzi OR, Lunin VG, Zalutsky MR, Sobolev AS. Engineered modular recombinant transporters: application of new platform for targeted radiotherapeutic agents to alpha-particle emitting 211 At. Int J Radiat Oncol Biol Phys. 2008;72(1):193–200.
40. DeVita Jr VT, Lawrence TS, Rosenberg SA, et al., editors. Cancer: principles & practice of oncology. 9th ed. Philadelphia, PA: LWW; 2011.
41. Puri RK, Hoon DS, Leland P, Snoy P, Rand RW, Pastan I, Kreitman RJ. Preclinical development of a recombinant toxin containing circularly permuted interleukin 4 and truncated Pseudomonas exotoxin for therapy of malignant astrocytoma. Cancer Res. 1996;56(24):5631–7.
42. Sampson JH, Reardon DA, Friedman AH, Friedman HS, Coleman RE, McLendon RE, Pastan I, Bigner DD. Sustained radiographic and clinical response in patient with bifrontal recurrent glioblastoma multiforme with intracerebral infusion of the recombinant targeted toxin TP-38: case study. Neuro Oncol. 2005;7(1):90–6.
43. vom Berg J, Vrohlings M, Haller S, Haimovici A, Kulig P, Sledzinska A, Weller M, Becher B. Intratumoral IL-12 combined with CTLA-4 blockade elicits T cell mediated glioma rejection. J Exp Med. 2013;210:2803–11.
44. Hayes RL, Arbit E, Odaimi M, et al. Adoptive cellular immunotherapy for the treatment of malignant gliomas. Crit Rev Oncol Hematol. 2001;39(1–2):31–42.
45. Grimm EA, Mazumder A, Zhang HZ, et al. Lymphokine-activated killer cell phenomenon. Lysis of natural killer-resistant fresh solid tumor cells by interleukin 2-activated autologous human peripheral blood lymphocytes. J Exp Med. 1982;155(6):1823–41.
46. Rosenberg SA. The development of new immunotherapies for the treatment of cancer using interleukin-2. A review. Ann Surg. 1988;208:121–35.
47. Lotze MT, Grimm EA, Mazumder A, Strausser JL, Rosenberg SA. Lysis of fresh and cultured autologous tumor by human lymphocytes cultured in T-cell growth factor. Cancer Res. 1981;41(11 Pt 1):4420–5.
48. Jacobs SK, Wilson DJ, Kornblith PL, Grimm EA. *In vitro* killing of human glioblastoma by interleukin-2-activated autologous lymphocytes. J Neurosurg. 1986;64(1):114–7.
49. Ahmed N, Salsman VS, Kew Y, et al. HER2-specific T cells target primary glioblastoma stem cells and induce regression of autologous experimental tumors. Clin Cancer Res. 2010;16(2):474–85.
50. Kahlon KS, Brown C, Cooper LJ, Raubitschek A, Forman SJ, Jensen MC. Specific recognition and killing of glioblastoma multiforme by interleukin 13-zetakine redirected cytolytic T cells. Cancer Res. 2004;64(24):9160–6.
51. Bullain SS, Sahin A, Szentirmai O, et al. Genetically engineered T cells to target EGFRvIII expressing glioblastoma. J Neurooncol. 2009;94(3):373–82.
52. Fedorov VD, Themeli M, Sadelain M. PD-1- and CTLA-4-based inhibitory chimeric antigen receptors (iCARs) divert off-target immunotherapy responses. Sci Transl Med. 2013;5:215ra172.
53. Steinman RM, Cohn ZAJ. J Exp Med. 1973;137:1142–62.
54. Badhiwala J, Decker WK, Berens ME, Bhardwaj RD. Clinical trials in cellular immunotherapy for brain/CNS tumors. Expert Rev Neurother. 2013;13(4):405–24.
55. Benencia F, Sprague L, McGinty J, Pate M, Muccioli M. Dendritic cells the tumor microenvironment and the challenges for an effective antitumor vaccination. J Biomed Biotechnol. 2012;2012:425–76.
56. Ali-Osman F. Brain tumors. In: Masters JRW, Palson B, editors. Human cell culture, vol. 2. New York, NY: Kluwer Academic Publishers; 1999. p. 167–84.
57. Hickey WF. Leukocyte traffic in the central nervous system: the participants and their roles. Semin Immunol. 1999;11(2):125–37.
58. Decker WK, Xing D, Shpall EJ. Dendritic cell immunotherapy for the treatment of neoplastic disease. Biol Blood Marrow Transplant. 2006;12(2):113–25.
59. Li Z, Lee JW, Mukherjee D, et al. Immunotherapy targeting glioma stem cells—insights and perspectives. Expert Opin Biol Ther. 2012;12(2):165–78.

60. Yamanaka R, Itoh K. Peptide-based immunotherapeutic approaches to glioma: a review. Expert Opin Biol Ther. 2007;7:645–9.
61. Heimberger AB, Crotty LE, Archer GE, et al. Epidermal growth factor receptor VIII peptide vaccination is efficacious against established intracerebral tumors. Clin Cancer Res. 2003;9(11):4247–54.
62. Debinski W, Gibo DM, Hulet SW, Connor JR, Gillespie GY. Receptor for interleukin 13 is a marker and therapeutic target for human high-grade gliomas. Clin Cancer Res. 1999;5(5):985–90.
63. Shimato S, Natsume A, Wakabayashi T, Tsujimura K, Nakahara N, Ishii J, Ito M, Akatsuka Y, Kuzushima K, Yoshida J. Identification of a human leukocyte antigen-A24-restricted T-cell epitope derived from interleukin-13 receptor alpha2 chain, a glioma-associated antigen. J Neurosurg. 2008;109(1):117–22.
64. Oka Y, Elisseeva OA, Tsuboi A, Ogawa H, Tamaki H, Li H, Oji Y, Kim EH, Soma T, Asada M, Ueda K, Maruya E, Saji H, Kishimoto T, Udaka K, Sugiyama H. Human cytotoxic T-lymphocyte responses specific for peptides of the wild-type Wilms' tumor gene (WT1) product. Immunogenetics. 2000;51(2):99–107.
65. Scheurer ME, Bondy ML, Aldape KD, Albrecht T, El-Zein R. Detection of human cytomegalovirus in different histological types of gliomas. Acta Neuropathol. 2008;116(1):79–86.
66. Okada H, Kalinski P, Ueda R, et al. Induction of CD8+ T-cell responses against novel glioma-associated antigen peptides and clinical activity by vaccinations with{α}-type 1 polarized dendritic cells and polyinosinic-polycytidylic acid stabilized by lysine and carboxymethylcellulose in patients with recurrent malignant glioma. J Clin Oncol. 2011;29(3):330–6.
67. Huang J, et al. Cytokine-induced killer (CIK) cells bound with anti-CD3/anti-CD133 bispecific antibodies target CD133(high) cancer stem cells in vitro and in vivo. Clin Immunol (Orlando, Fla). 2013;149(1):156–68.
68. Ikegame A, et al. Small molecule antibody targeting HLA class I inhibits myeloma cancer stem cells by repressing pluripotency-associated transcription factors. Leukemia. 2012;26(9):2124–34.
69. Schlaak M, et al. Regression of metastatic melanoma in a patient by antibody targeting of cancer stem cells. Oncotarget. 2012;3(1):22–30.
70. Ueda R, et al. Identification of HLA-A2- and A24-restricted T-cell epitopes derived from SOX6 expressed in glioma stem cells for immunotherapy. Int J Cancer. 2010; 126(4):919–29.
71. Xu Q, et al. Antigen-specific T-cell response from dendritic cell vaccination using cancer stem-like cell-associated antigens. Stem Cells (Dayton, Ohio). 2009;27(8):1734–40.
72. Xiao ZY, et al. An experimental study of dendritic cells transfected with cancer stem-like cells RNA against 9L brain tumors. Cancer Biol Ther. 2011;11(11):974–80. Web. 21 Mar. 2014.
73. Pellegatta S, et al. Neurospheres enriched in cancer stem-like cells are highly effective in eliciting a dendritic cell-mediated immune response against malignant gliomas. Cancer Res. 2006;66(21):10247–52.
74. Ning N, et al. Cancer stem cell vaccination confers significant antitumor immunity. Cancer Res. 2012;72(7):1853–64. Web. 21 Mar. 2014.
75. Natsume A, et al. The DNA demethylating agent 5-aza-2′-deoxycytidine activates NY-ESO-1 antigenicity in orthotopic human glioma. Int J Cancer. 2008;122(11):2542–53. Web. 21 Mar. 2014.
76. Odunsi K, et al. NY-ESO-1 and LAGE-1 cancer-testis antigens are potential targets for immunotherapy in epithelial ovarian cancer. Cancer Res. 2003;63(18):6076–83.
77. Jacobs SK, Wilson DJ, Kornblith PL, Grimm EA. Interleukin-2 or autologous lymphokine-activated killer cell treatment of malignant glioma: phase I trial. Cancer Res. 1986;46(4 Pt 2):2101–4.
78. Barba D, Saris SC, Holder C, Rosenberg SA, Oldfield EH. Intratumoral LAK cell and interleukin-2 therapy of human gliomas. J Neurosurg. 1989;70(2):175–82.

79. Blancher A, Roubinet F, Grancher AS, et al. Local immunotherapy of recurrent glioblastoma multiforme by intracerebral perfusion of interleukin-2 and LAK cells. Eur Cytokine Netw. 1993;4(5):331–41.
80. Boiardi A, Silvani A, Ruffini PA, et al. Loco-regional immunotherapy with recombinant interleukin-2 and adherent lymphokine-activated killer cells (A-LAK) in recurrent glioblastoma patients. Cancer Immunol Immunother. 1994;39(3):193–7.
81. Merchant RE, Grant AJ, Merchant LH, Young HF. Adoptive immunotherapy for recurrent glioblastoma multiforme using lymphokine activated killer cells and recombinant interleukin-2. Cancer. 1988;62(4):665–71.
82. Merchant RE, Merchant LH, Cook SH, McVicar DW, Young HF. Intralesional infusion of lymphokine-activated killer (LAK) cells and recombinant interleukin-2 (rIL-2) for the treatment of patients with malignant brain tumor. Neurosurgery. 1988;23(6):725–32.
83. Sankhla SK, Nadkarni JS, Bhagwati SN. Adoptive immunotherapy using lymphokine-activated killer (LAK) cells and interleukin-2 for recurrent malignant primary brain tumors. J Neurooncol. 1996;27(2):133–40.
84. Bielamowicz K, Khawja S, Ahmed N. Adoptive cell therapies for glioblastoma. Front Oncol. 2013;3:275.
85. Yoshida S, Tanaka R, Takai N, Ono K. Local administration of autologous lymphokine-activated killer cells and recombinant interleukin 2 to patients with malignant brain tumors. Cancer Res. 1988;48(17):5011–6.
86. Okamoto Y, Shimizu K, Tamura K, et al. An adoptive immunotherapy of patients with medulloblastoma by lymphokine-activated killer cells (LAK). Acta Neurochir. 1988;94(1–2): 47–52.
87. Dillman RO, Duma CM, Schiltz PM, et al. Intracavitary placement of autologous lymphokine-activated killer (LAK) cells after resection of recurrent glioblastoma. J Immunother. 2004;27(5):398–404.
88. Dillman RO, Duma CM, Ellis RA, et al. Intralesional lymphokine-activated killer cells as adjuvant therapy for primary glioblastoma. J Immunother. 2009;32(9):914–9.
89. Kitahara T, Watanabe O, Yamaura A, et al. Establishment of interleukin 2 dependent cytotoxic T lymphocyte cell line specific for autologous brain tumor and its intracranial administration for therapy of the tumor. J Neurooncol. 1987;4(4):329–36.
90. Kruse CA, et al. Treatment of recurrent glioma with intracavitary alloreactive cytotoxic T lymphocytes and interleukin-2. Cancer Immunol Immunother. 1997;45(2):77–87.
91. Tsuboi K, Saijo K, Ishikawa E, et al. Effects of local injection of ex vivo expanded autologous tumor-specific T lymphocytes in cases with recurrent malignant gliomas. Clin Cancer Res. 2003;9(9):3294–302.
92. Tsurushima H, Liu SQ, Tuboi K, et al. Reduction of end-stage malignant glioma by injection with autologous cytotoxic T lymphocytes. Jpn J Cancer Res. 1999;90(5):536–45.
93. Quattrocchi KB, Miller CH, Cush S, et al. Pilot study of local autologous tumor infiltrating lymphocytes for the treatment of recurrent malignant gliomas. J Neurooncol. 1999;45(2):141–57.
94. Holladay FP, Heitz-Turner T, Bayer WL, Wood GW. Autologous tumor cell vaccination combined with adoptive cellular immunotherapy in patients with Grade III/ IV astrocytoma. J Neurooncol. 1996;27(2):179–89.
95. Plautz GE, Barnett GH, Miller DW, et al. Systemic T cell adoptive immunotherapy of malignant gliomas. J Neurosurg. 1998;89(1):42–51.
96. Plautz GE, Miller DW, Barnett GH, et al. T cell adoptive immunotherapy of newly diagnosed gliomas. Clin Cancer Res. 2000;6(6):2209–18.
97. Sloan AE, Dansey R, Zamorano L, et al. Adoptive immunotherapy in patients with recurrent malignant glioma: preliminary results of using autologous whole-tumor vaccine plus granulocyte–macrophage colony-stimulating factor and adoptive transfer of anti-CD3-activated lymphocytes. Neurosurg Focus. 2000;9(6):e9.
98. Wood GW, Holladay FP, Turner T, Wang YY, Chiga M. A pilot study of autologous cancer cell vaccination and cellular immunotherapy using anti-CD3 stimulated lymphocytes in patients with recurrent Grade III/IV astrocytoma. J Neurooncol. 2000;48(2):113–20.

99. Kronik N, Kogan Y, Vainstein V, Agur Z. Improving alloreactive CTL immunotherapy for malignant gliomas using a simulation model of their interactive dynamics. Cancer Immunol Immunother. 2008;57(3):425–39.
100. Mitchell DA, Xie W, Schmittling R, et al. Sensitive detection of human cytomegalovirus in tumors and peripheral blood of patients diagnosed with glioblastoma. Neuro Oncol. 2008;10(1):10–8.
101. Cobbs CS, Harkins L, Samanta M, et al. Human cytomegalovirus infection and expression in human malignant glioma. Cancer Res. 2002;62(12):3347–50.
102. Dziurzynski K, Wei J, Qiao W, et al. Glioma-associated cytomegalovirus mediates subversion of the monocyte lineage to a tumor propagating phenotype. Clin Cancer Res. 2011;17(14):4642–9.
103. Ishikawa E, Tsuboi K, Yamamoto T, et al. Clinical trial of autologous formalin-fixed tumor vaccine for glioblastoma multiforme patients. Cancer Sci. 2007;98(8):1226–33.
104. Schneider T, Gerhards R, Kirches E, Firsching R. Preliminary results of active specific immunization with modified tumor cell vaccine in glioblastoma multiforme. J Neurooncol. 2001;53(1):39–46.
105. Steiner HH, Bonsanto MM, Beckhove P, et al. Antitumor vaccination of patients with glioblastoma multiforme: a pilot study to assess feasibility, safety, and clinical benefit. J Clin Oncol. 2004;22(21):4272–81.
106. Kikuchi T, Akasaki Y, Irie M, Homma S, Abe T, Ohno T. Results of a Phase I clinical trial of vaccination of glioma patients with fusions of dendritic and glioma cells. Cancer Immunol Immunother. 2001;50(7):337–44.
107. Yu JS, et al. Vaccination of malignant glioma patients with peptide-pulsed dendritic cells elicits systemic cytotoxicity and intracranial t-cell infiltration. Cancer Res. 2001;61(3): 842–7.
108. Yu JS, Liu G, Ying H, Yong WH, Black KL, Wheeler CJ. Vaccination with tumor lysate-pulsed dendritic cells elicits antigen-specific, cytotoxic T-cells in patients with malignant glioma. Cancer Res. 2004;64(14):4973–9.
109. Wheeler CJ, Black KL, Liu G, et al. Vaccination elicits correlated immune and clinical responses in glioblastoma multiforme patients. Cancer Res. 2008;68(14):5955–64.
110. Fadul CE, et al. Immune response in patients with newly diagnosed glioblastoma multiforme treated with intranodal autologous tumor lysate-dendritic cell vaccination after radiation chemotherapy. J Immunother (Hagerstown, Md 1997). 2011;34(4):382–9.
111. Jie X, Hua L, Jiang W, Feng F, Feng G, Hua Z. Clinical application of a dendritic cell vaccine raised against heat-shocked glioblastoma. Cell Biochem Biophys. 2012;62(1):91–9.
112. Cho DY, Yang WK, Lee HC, et al. Adjuvant immunotherapy with whole-cell lysate dendritic cells vaccine for glioblastoma multiforme: a Phase II clinical trial. World Neurosurg. 2012;77(5–6):736–44.
113. Westphal M, Ferdinand B. Final results of a randomized phase III trial of nimotuzumab for the treatment of newly diagnosed glioblastoma in addition to standard radiation and chemotherapy with temozolomide versus standard radiation and temoziolamide. J Clin Oncol. 2012;30(Suppl):abstr 2033. Web. 28 July 2014.
114. Wen PY, et al. A randomized, double-blind, placebo-controlled phase 2 trial of dendritic cell (DC) vaccination with ICT-107 in newly diagnosed glioblastoma (GBM) patients. J Clin Oncol. 2014;32(5s Suppl):abstr 2005. Web. 28 July 2014.
115. Yajima N, Yamanaka R, Mine T, et al. Immunologic evaluation of personalized peptide vaccination for patients with advanced malignant glioma. Clin Cancer Res. 2005; 11(16):5900–11.
116. Izumoto S, Tsuboi A, Oka Y, et al. Phase II clinical trial of Wilms tumor 1 peptide vaccination for patients with recurrent glioblastoma multiforme. J Neurosurg. 2008;108(5):963–71.
117. Morita S, Oka Y, Tsuboi A, et al. A Phase I/II trial of a WT1 (Wilms' tumor gene) peptide vaccine in patients with solid malignancy: safety assessment based on the Phase I data. Jpn J Clin Oncol. 2006;36(4):231–6.

118. Sampson JH, Aldape KD, Archer GE, et al. Greater chemotherapy-induced lymphopenia enhances tumor-specific immune responses that eliminate EGFRvIII-expressing tumor cells in patients with glioblastoma. Neuro Oncol. 2011;13(3):324–33.
119. Sampson JH, Heimberger AB, Archer GE, et al. Immunologic escape after prolonged progression-free survival with epidermal growth factor receptor variant III peptide vaccination in patients with newly diagnosed glioblastoma. J Clin Oncol. 2010;28(31):4722–9.
120. Crane CA, Han SJ, Ahn B, et al. Individual patient-specific immunity against high-grade glioma after vaccination with autologous tumor derived peptides bound to the 96 KD chaperone protein. Clin Cancer Res. 2013;19(1):205–14.
121. Vik-Mo EO, Nyakas M, Mikkelsen BV, Moe MC, Due-Tønnesen P, Suso EM, Sæbøe-Larssen S, Sandberg C, Brinchmann JE, Helseth E, Rasmussen AM, Lote K, Aamdal S, Gaudernack G, Kvalheim G, Langmoen IA. Therapeutic vaccination against autologous cancer stem cells with mRNA-transfected dendritic cells in patients with glioblastoma. Cancer Immunol Immunother. 2013;62(9):1499–509.
122. Bigner DD, Brown M, Coleman RE, et al. Phase I studies of treatment of malignant gliomas and neoplastic meningitis with 131I-radiolabeled monoclonal antibodies anti-tenascin 81C6 and anti-chondroitin proteoglycan sulfate Me1-14F(ab′)2—a preliminary report. J Neurooncol. 1995;24(1):109–22.
123. Kurpad SN, Zhao XG, Wikstrand CJ, Batra SK, McLendon RE, Bigner DD. Tumor antigens in astrocytic gliomas. Glia. 1995;15(3):244–56.
124. Murphy-Ullrich JE, Lightner VA, Aukhil I, Yan YZ, Erickson HP, Hook M. Focal adhesion integrity is downregulated by the alternatively spliced domain of human tenascin. J Cell Biol. 1991;115(4):1127–36.
125. Neyns B, Sadones J, Joosens E, et al. Stratified phase II trial of cetuximab in patients with recurrent high-grade glioma. Ann Oncol. 2009;20(9):1596–603.
126. Gil-gil JM, et al. Bevacizumab for the treatment of glioblastoma. Clin Med Insights Oncol. 2013;7:123–35.
127. de Groot JF, Yung WK. Bevacizumab and irinotecan in the treatment of recurrent malignant gliomas. Cancer J. 2008;14(5):279–85.
128. Lai A, Tran A, Nghiemphu PL, et al. Phase II study of bevacizumab plus temozolomide during and after radiation therapy for patients with newly diagnosed glioblastoma multiforme. J Clin Oncol. 2011;29:142–8.
129. Gilbert MR, et al. RTOG 0825: Phase III double-blind placebo-controlled trial evaluating bevacizumab (Bev) in patients (Pts) with newly diagnosed glioblastoma (GBM). J Clin Oncol. 2013;31.

Emerging Strategies for the Treatment of Tumor Stem Cells in Central Nervous System Malignancies

Imad Saeed Khan and Moneeb Ehtesham

Abstract High-grade central nervous system (CNS) tumors are notorious for high rates of recurrence and poor outcomes. A small cohort of tumor cells, dubbed tumor stem cells (TSC), are now being recognized as an important subset of the tumor that is resistant to chemotherapy and radiotherapy and account for the high recurrence rates. Recent research is developing modalities to target TSCs specifically in a bid to improve the response of the tumor as a whole. The methods being employed to target TSCs include targeting TSC-specific pathways or receptors, TSC-sensitizing agents to chemotherapy and radiotherapy, immunotherapy, TSC-differentiating agents, and viral therapy. This chapter provides an overview of strategies that are expected to help develop new and more effective treatments for CNS tumors.

Keywords Glioma stem cells • Tumor stem cells • Cancer stem cells • Chemotherapy sensitization • Radiotherapy sensitization • Immunotherapy • Differentiation agents • Virotherapy • Gene therapy

Introduction

Central nervous system (CNS) tumors are notorious for including some of the most lethal tumors in humans. The most common intrinsic brain tumor, the glioblastoma multiforme (GBM), carries a uniformly poor prognosis with most patients not surviving up till 2 years after diagnosis. The standard management strategy for patients with GBM is based on the protocol described by Stupp and

I.S. Khan, M.D.
Section of Neurosurgery, Geisel School of Medicine at Dartmouth, Lebanon, NH, USA

M. Ehtesham, M.D. (✉)
Department of Neurological Surgery, Vanderbilt University Medical Center, Nashville, TN, USA
e-mail: moneeb.ehtesham@Vanderbilt.Edu

M. Ehtesham (ed.), *Stem Cell Biology in Neoplasms of the Central Nervous System*, Advances in Experimental Medicine and Biology 853,
DOI 10.1007/978-3-319-16537-0_9

colleagues: specifically, maximal safe surgical excision followed by radiotherapy and temozolomide (TMZ) chemotherapy [1]. Unfortunately, in spite of these aggressive measures, recurrence almost always occurs. This therapeutic regimen has only been able to increase the median survival for GBM from 12.1 months to the current 14.6 months [1]. The current prognosis of the disease stresses the importance of developing novel treatment strategies and therapeutics targeting tumor stem cell (TSC) populations have recently received notable attention in this regard.

The TSC hypothesis is based upon the presence of a small subset of tumor cells with properties akin to stem cells. According to this premise, TSCs sit at the apex of all tumor cells and exhibit properties of multi-lineage capacity and self-renewal [2]. While self-renewal maintains the population of the TSCs, the process of differentiation produces downstream tumor progenitor cells that generate the genetically diverse progeny of the tumor mass.

An important property of TSCs is the ability to initiate tumors when xenografted in nude mice. The xenograft initiation efficiency is significantly higher than implantation of traditional GBM cell lines [3, 4]. Additionally, TSCs are generally more resistant to conventional cytotoxic therapy, leading to tumor repopulation via differentiation of unaffected TSCs after cytotoxic therapy. Therefore, TSCs are thought to be a major factor driving recurrence and therapeutic resistance in gliomas (Fig. 1).

Challenges with Current Treatment Strategies

Current therapeutic strategies advocate a uniform regimen for patients with CNS tumors. For chemotherapy in GBM, TMZ is considered an essential part of the treatment approach. TMZ causes cytotoxicity against GBM by the creation of O6-methylguanine (O6MeG) lesions—leading to DNA fragmentation and disruption of DNA replication. The resulting effects include tumor suppression and tumor cell apoptotic cell death [5].

While the addition of TMZ to the chemotherapy protocol is only able to improve the median survival to 14.6 months, Heigi and colleagues reported a specific patient cohort of long-term GBM survivors with a median survival of 21.7 months [6]. Further investigation of their cohort revealed an absence of tumor methylguanine-DNA methyltransferase (MGMT) expression in their patients [6]. By removing the methyl groups added on by TMZ, MGMT prevents tumor cell death. However, methylation of its promoter leads to absent or reduced expression of the MGMT and increases the cytotoxic efficacy of TMZ. The overall effect is that of increased tumor cell death, translating into improved patient survival.

The effect of MGMT status on the response to treatment points towards the importance of understanding the differences within the tumor cell cohort that

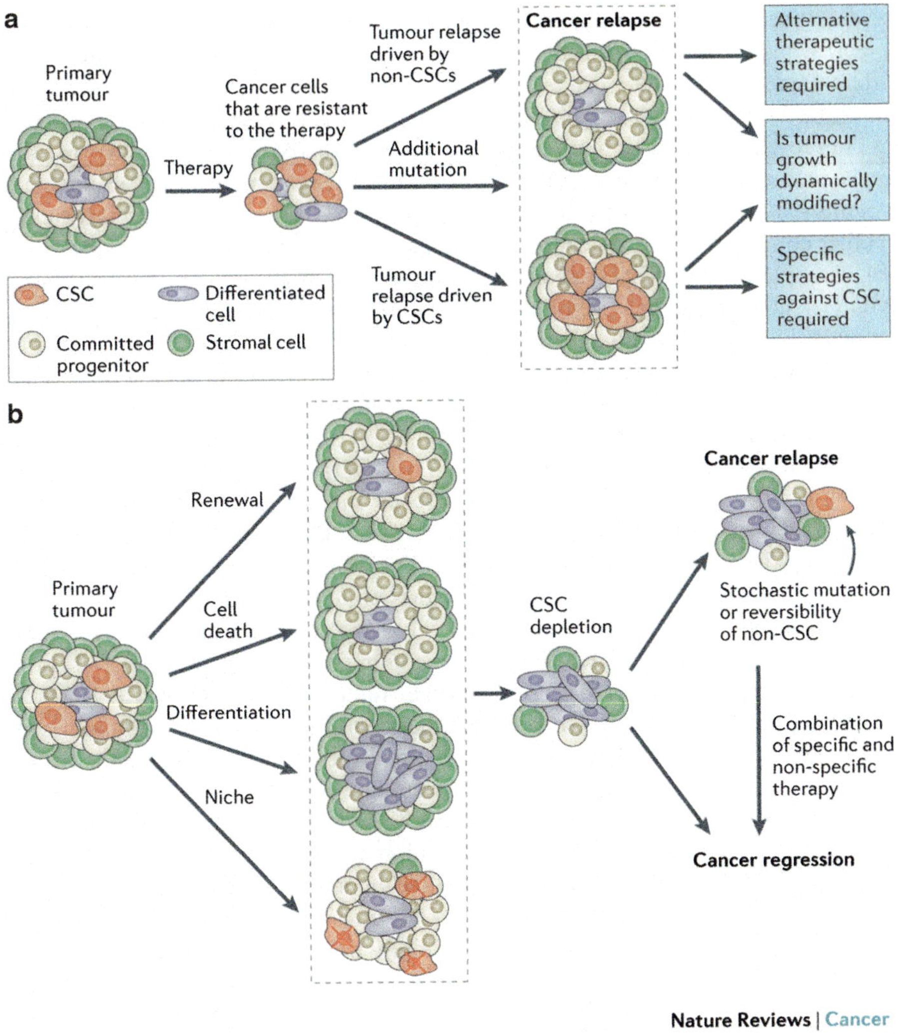

Fig. 1 Implication of cancer stem cells (CSCs) in cancer therapies and tumor relapse. (**a**) Anticancer therapies may not kill all tumor cells equally. CSCs that sustain tumor growth or another population of more slowly cycling tumor cells may be responsible for tumor resistance to therapies and tumor relapse. Depending on the population responsible for tumor relapse, new strategies should be designed to eradicate all tumor cells. (**b**) The CSC model suggests that inhibiting CSC renewal or promoting their differentiation should induce tumor regression. Drugs could impair CSC self-renewal, induce their specific cell death, induce their differentiation, or target their niche. All of these strategies would lead to the depletion of the pool of CSCs and subsequent tumor regression. However, if the CSC potential is reversible, or if newly acquired mutations confer resistance to therapy, then tumor regression would only be transient, leading to cancer relapse (reprinted with permission from Beck B, Blanpain C, Nat Rev Cancer. 2013 Oct;13(10):727–38. Unraveling cancer stem cell potential)

dictates the ultimate response to treatment. For treatment purposes the TSC fraction is increasingly being recognized as an important, and in some ways fundamentally different, part of the tumor. Liu and colleagues reported that CD133+ cells depicted a multifold higher activity of MGMT compared to CD133− cells, which translates into improved DNA repair and increased resistance to TMZ [7, 8]. Another reason for the increased resistance to TMZ may be the downregulation of autophagy-related proteins in the TSCs [9]. TSCs have also shown to possess stronger drug resistance to other conventional anticancer drugs, such as doxorubicin (Dox), etoposide (VP-16), carboplatin, and BCNU due to an enhanced expression of multidrug resistance (MDR) 1 [10]. Thus, increasing evidence points towards the relatively refractory nature of TSCs to conventional chemotherapy.

While Beier and colleagues were able to show that TMZ induced a dose- and time-dependent decline of brain TSCs in a cell culture study, TMZ needed clinically unreachable levels to be effective [11]. Glioma TSCs also show an upregulation of mRNAs of FAS-associating death domain (FADD)-like antiapoptotic molecule (FLIP), B-cell CLL/lymphoma 2 (Bcl-2), Bcl-X, and some inhibitor of apoptosis (IAP) family members [12–14]. Other factors that confer a protective advantage to TSCs include a higher expression of breakpoint cluster region pseudogene 1 (BCRP1; drug-resistant gene) and antiapoptosis proteins and inhibitors [7].

The fraction of tumor cells expressing CD133 is also known to be enriched after radiation in gliomas [15]. CD133-expressing glioma cells survive ionizing radiation in increased proportions relative to most other tumor cells. This is because TSCs preferentially activate the DNA damage checkpoint in response to radiation, and repair radiation-induced DNA damage more effectively than CD133-negative tumor cells. With exposure to conventional radiation, CD133+ cells exhibit enhanced activation of three key mediators of cell cycle check points: Rad17, Chk1, and Chk2 [16, 17]. Interestingly, if administered specific inhibitors of the Chk1 and Chk2 checkpoint kinases TSCs become more radiosensitive, akin to CD133− tumor cells [16].

Due to their inherent resistant nature, TSCs are worthwhile targets for the development of specific treatment modalities to improve the overall response of tumors to treatment [18]. Targeting a specific molecular protein signal pathway of TSCs with a therapeutic target is one of the ways investigators are aiming to eradicate these cells. Other strategies include virotherapy, increasing TSC chemosensitivity and radiosensitivity by using hypersensitivity agents [19, 20], immunotherapy using autologous dendritic cells, and using differentiation agents in a bid to promote differentiation of TSCs [21]. Improving knowledge of the unique characteristics of TSCs is driving the development of TSC-specific therapeutics. Based on the suggested pivotal role of TSCs in the origin, development, and maintenance of tumors, future therapies will aim to effectively eradicate them to improve the response rates in tumors and decrease recurrences. We will now review some of the basic strategies being employed to target TSCs that are expected to help engineer more effective treatment strategies in the future.

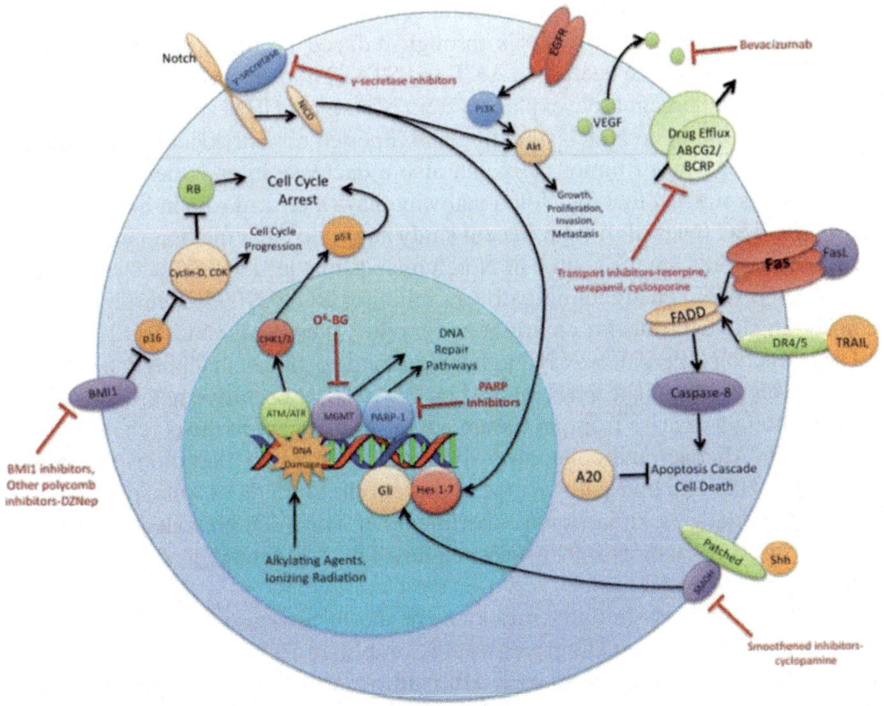

Fig. 2 Mediators of TSC treatment resistance. Depicted are the various treatment resistance mechanisms and pathways differentially expressed or regulated in TSC versus their differentiated cell counterparts. *Blocked red lines* indicate ways to inhibit or block these mediators (from Schmalz PG1, Shen MJ, Park JK. Cancers (Basel). 2011 Feb 10;3(1):621–35. Treatment resistance mechanisms of malignant glioma tumor stem cells (open access))

Targeting TSC-Specific Pathways and Receptors

One of the major methods to target TSCs is to identify pathways and/or receptors that are specific for TSCs (Fig. 2). These pathways can then be exploited to decrease the number of TSCs while combining with conventional therapeutics will treat the overall tumor mass. Some of the major targets of interest are summarized below.

Notch ligands, receptors, and targets have been found in a wide range of neoplasms, including, but not limited to, lung, breast, cervix, renal, pancreas, medulloblastoma (MB), and GBM [22–31]. Additionally, in many of these tumors increased Notch activity has been shown to promote tumor growth, with studies showing that Notch pathway blockade inhibits proliferation of tumor cells. In the CNS, Notch signaling pathway regulates neural stem cells (NSCs). Studies have also demonstrated higher Notch activity in CNS TSCs [32].

The Notch pathway blockade by gamma-secretase inhibitors is another important pathway that depletes glioma TSCs through reduced proliferation and increased apoptosis associated with decreased AKT and STAT3 phosphorylation [33]. Using a three-dimensional organotypic explant system of surgical GBM specimens, Hovinga and colleagues inhibited Notch signaling and reported not only decreased proliferation and self-renewal of tumor cells, but also a decrease in endothelial cells [25]. These findings suggest that the Notch pathway plays a critical role in linking angiogenesis and TSC renewal. A more recent study suggested that the brain microvascular endothelial cells are the source of Notch ligands that lead to TSC sustenance and renewal [31]. A Notch signaling pathway inhibitor RO4929097 is currently being evaluated in clinical trials for recurrent and progressive GBMs (NCT01122901).

The hedgehog (Hh) pathway is another significant pathway that plays an essential role in development of the cerebellum [34, 35]. MB, a primitive neuroectodermal tumor, is thought to arise from immature neural progenitors in the cerebellum [36]. Additionally, Michael and colleagues showed that genomic alterations in components of the Hh signaling pathway were present up to 25 % of human MBs [37]. Additional work using knockdown experiments of Bmi1 demonstrated that Hh signaling drives Bmi1 expression, which is a key TSC regulatory gene implicated in the pathogenesis of MB [38].

The Hh pathway is similarly important in the pathogenesis of gliomas. Gli, a component of the Hh signaling pathway, is amplified in gliomas [39]. Bar et al. reported that cyclopamine blocks the Hh pathway causing a depletion of TSC in GBM [40]. Likewise, Clement and colleagues reported that interference of Hh-Gli signaling with cyclopamine or through lentiviral mediated silencing resulted in decreased self-renewal and tumorigenicity of TSCs [41]. SANT-1 inhibition of Hh has also been shown to reduce proliferation of glioma TSCs [42].

Glioma TSCs have also shown a positive correlation with microvessel density and have multiple regulatory roles in endothelial cells [43]. They are thought to enhance the migration and proliferation of the endothelial cells by secretion of sonic hedgehog (Shh), leading to activation of the Hh pathway of the endothelial cells [44]. Consequently, GDC-0449 or vismodegib (a small-molecule antagonist of the Hh pathway) has recently garnered interest and is being tested in a clinical trial for recurrent GBM (NCT00980343) [45]. A case report of a patient with refractory metastatic MB managed with GDC-0449 has also been reported. This treatment resulted in rapid (although transient) regression of the tumor and reduction of symptoms [46]. GDC-0449 is also being evaluated in clinical trials for recurrent and recalcitrant MB (NCT00939484 and NCT00822458).

Bao and colleagues were the first to report the intimate relationship between glioma TSCs and the microvasculature. They reported that CD133+ cells produced high levels of VEGF that induced endothelial cell migration. Conversely, treatment of CD133+ GBM cells with bevacizumab blocked the tumor cells' ability to induce endothelial cell migration and initiate tumors in vivo [47]. Similarly, Calabrese et al. demonstrated that treatment of GBM with bevacizumab depleted tumor blood vessels and caused a significant reduction in the number of GBM TSCs [48]. Due to the efficacy of antiangiogenic agents in preclinical studies, they have been tested in clinical trials. Unfortunately, bevacizumab has failed to show improvement in the overall survival of patients with newly diagnosed GBM [49]. Similarly, cediranib did not prolong progression-free survival in patients with recurrent GBM, either as monotherapy or in combination with lomustine, compared to patients who were treated with lomustine alone [50].

Amplification and/or mutation of receptor tyrosine kinases, such as epidermal growth factor receptor (EGFR), is another common genetic alteration in GBM [51, 52]. Recent studies have demonstrated the presence of a constitutively active EGFR mutant (EGFRvIII) associated with glioma TSCs. This pathway potentiates tumor growth and heterogeneity through IL-6-mediated Notch signaling [53, 54], and Src family kinase (SFK)-dependent phosphorylation of Dock180 [55, 56]. Clinical trails investigating the efficacy of EGFR inhibitors however have yielded disappointing results [57–59].

Aberrant Wnt signaling is molecularly linked to many human cancers, including colorectal, breast, ovarian, and hepatocellular carcinoma, neuroectodermal tumors, and glioma [60–63]. Dysregulation of the Wnt-pathway has also been documented in glioma TSCs [64, 65]. Investigators have also identified a role for this pathway in MBs [66, 67]. Other similar targets currently being investigated to treat CNS TSCs include the homeobox (HOx) family [7], phosphatase and tensin (PTEN) [68], telomerase [69], efflux transporters [70, 71], and microRNA [72–74].

Chemotherapy and Radiotherapy Sensitizers

Increased resistance to chemotherapy is a great challenge when treating TSCs. However investigators have reported several ways to potentiate the cytotoxicity of chemotherapeutic agents. These include cell-cycle checkpoint abrogation [75, 76], depletion in the expression of antiapoptosis proteins [77], and DNA repair enzymes [78].

A molecular chaperone, 90-kDa heat-shock protein (hsp90), has recently been described as a chemotherapy sensitizer because it is expressed at 2–10-fold higher levels in tumors compared to normal tissues [79]. Ohba and colleagues reported that inhibition of hsp90 potentiated the cytotoxicity of chemotherapeutic agents in human glioma cell lines [80]. On the other hand, Sauvageot and colleagues reported that while 17-AAG (inhibitor of hsp90) inhibited the growth of glioma cells and although it has a synergistic effect with radiation, it was not found to synergize with TMZ [81].

GPI 15427, a novel poly(ADP-ribose) polymerase-1 (PARP-1) inhibitor, significantly increases the life span of its tumor-bearing mice when it is administered systemically shortly before TMZ [82]. The same group later used the oral route to administer GPI 15427 and found it to be efficacious as a chemosensitizer as well [83, 84].

More recently, the effect of secreted frizzled-related protein 4 (sFRP4), a Wnt signaling antagonist, in chemosensitizing glioma TSCs was examined. The results indicated that sFRP4 was able to significantly sensitize glioma TSCs to doxorubicin or cisplatin [85]. Similarly, another study used proteasome inhibitor bortezomib and revealed that combination therapies based on bortezomib and bevacizumab offered an increased benefit when the two agents are used in combination [86]. Xu et al. targeted CD44, which is upregulated in GBM, and reported that its depletion impeded the growth of GBM and sensitized the tumor cells to cytotoxic drugs in vivo [87]. Tyrosine kinase inhibitors have also been experimented for sensitization of the tumor. Wachsberger and colleagues used cediranib, a potent receptor tyrosine kinase inhibitor that inhibits all three VEGF receptors. They reported that while cediranib did not radiosensitize the glioma cells, it did enhance the effectiveness of TMZ [88].

1,3-Bis(2-chloroethyl)-1-nitrosourea (BCNU) is one of the most commonly used chemotherapeutic agents in the treatment of GBM but it often fails to eradicate TSCs. Research has uncovered an overexpression of multiple ion channel genes that are related to drug efflux. However when a chloride channel blocker, 4,4′-diisothiocyanostilbene-2,2′-disulfonic acid, is used in combination with BCNU, the effect of BCNU is seen to synergistically increase [89].

Some investigators have aimed to disrupt the TSC niche by administering antiangiogenic agents with the intention of disrupting the stemness of the tumor cells. By losing the stem-cell characteristics the tumor cells may become more sensitized to chemotherapy. This technique has been used by researchers to show that combined antiangiogenic and cytotoxic drugs can result in a significant reduction in the number of glioma TSCs [90].

There has also been a concentrated effort to understand the biology of TSC radioresistance and develop approaches to sensitize the tumor cells to ionizing radiation. As TGF-β is a modifier of radiation responses, TGF-β receptor (TGFβR) I kinase inhibitor (LY2109761) has been used in combination with radiotherapy as an approach to increase the radiosensitivity of glioma cell lines including in TSCs [91]. Similarly, LY364947, another small-molecule inhibitor of TGF-β type I receptor kinase, was used by investigators to show improved tumor response when it was administered prior to radiotherapy [92]. A TGF-β inhibitor, LY2157299, alongside TMZ-based treatment regimen is also being evaluated in an ongoing clinical trial (NCT01220271).

EGFR activation has also been implicated in the radioresistance of many cancers, including brain tumors. Combining EGFR targeting with radiotherapy is an appealing option to increase the cytotoxic effect of radiation. To test this strategy, Geoerger et al. used gefitinib (tyrosine kinase inhibitor) in two xenograft models: an EGFR-amplified glioma and an EGFR-expressing ependymoma. For both the models, there was a positive trend towards superior antitumor activity when combined therapy was administered (gefitinib + radiation) [93].

Kang and colleagues further investigated the effect of gefitinib in glioma TSCs and found that it enhanced radiosensitivity of TSCs by reducing EGFR-Akt activation and DNA-PKcs expression. This was accompanied by enhanced irradiation-induced DNA double-strand breaks and inhibition of its repair [94]. Likewise, another group investigated the efficacy of ZD1839 (Iressa), a selective EGFR tyrosine kinase inhibitor, on the radiation sensitivity of the U251 GBM cell line. In their radiation survival experiments, ZD1839 had a significant radiosensitizing effect and increased tumor cell death [95]. In the clinical domain, a phase 1/2 study of radiation therapy with concurrent gefitinib for newly diagnosed GBM showed good tolerance of the drug but no benefit in survival [96]. Other tyrosine kinase inhibitors investigated as radiosensitizers for GBM include erlotinib [57] and vandetanib [97].

Signal transducer and activator of transcription (STAT) 3 is a member of a family of DNA-binding molecules, and the aberrant activity of the JAK2/STAT3 pathway is associated with glioma TSCs. Inhibition of this pathway leads to decreased proliferation of glioma TSCs [98, 99]. Yang and colleagues reported that resveratrol (inhibitor of the STAT3 axis) therapy significantly improved the survival rate in their xenotransplant model in part by synergistically enhancing the radiosensitivity of radiation-treated GBM TSCs [100].

STAT3 pathway also plays a key role in mediating CSC properties in MB-derived CD133(+) cells [101]. Celecoxib is a selective COX-2 inhibitor and has been shown to potentially reduce STAT3 phosphorylation [102, 103]. Incubation of MB TSCs with celecoxib has shown to dose-dependently suppress the TSC properties of the tumor cells and enhance the radiotherapy effect on the induction of apoptosis [104]. Similarly, inhibition of phosphorylated STAT3 by cucurbitacin I has also demonstrated enhancement of the chemoradiosensitivity of MB TSCs [101].

Valproic acid (VPA) is a commonly prescribed antiepileptic drug used for the management of seizures in brain tumor patients. Besides its antiseizure property, VPA is an effective inhibitor of histone deacetylase and is involved in modulating chromatin structure and gene expression [105–107]. Interaction between VPA and TMZ has been studied to depict enhanced cytotoxicity in TMZ-sensitive cell line (D384) and the TMZ-resistant cell line (T98). The enhancement of TMZ-induced apoptosis is associated with increased reactive oxygen species production and glutathione depletion. Pretreatment with *N*-acetylcysteine can partially recover the apoptotic effect of the TMZ/VPA combination treatment [108]. Furthermore, the combination of VPA and TMZ also causes significant radiation enhancement in the glioma cell lines [109].

Another approach to make glioma TSCs more radiosensitive is to inhibit the DNA damage responses (DDR) that follow radiotherapy [110]. A dual phosphoinositide 3-kinase/mTOR inhibitor NVP-BEZ235 can potently inhibit two central DDR kinases, DNA-dependent protein kinase catalytic subunit (DNA-PKcs) and ataxia-telangiectasia mutated (ATM), and has been shown to potentiate the damage caused by ionizing radiation in glioma cells [111, 112]. The recognition of various pathways and receptors that can be modulated to increase the chemoradiosensitivity of CNS TSCs is an area of intense research that promises to identify specific clinical targets that may be exploited.

Immunotherapy

GBMs secrete multiple immunosuppressive factors, including transforming growth factor-β (TGF-β) and prostaglandin E2 (PGE2), which lead to a profound immunosuppressive effect both locally and systemically [113, 114]. TGF-β expands the pool of immunosuppressive regulatory T cells, resulting in suppression of T cell proliferation. Additionally, TGF-β and PGE2 downregulate the expression of major histocompatibility complex (MHC) class II, as well as the antigen processing of dendritic cells (DCs) [115]. Disruption of the immunosuppressive environment represents a promising immunological target to treat tumor cells.

Glioma cells also express certain antigens that are not expressed elsewhere in the brain. These antigens can be recognized by T cells, which can play an important role in tumor rejection. Some of the major glioma antigens include MAGE-1 [116], SOX6 [117], gp100, TRP-2 [118, 119], EGFRvIII [120], L13Ra2 [121], HER-2 [118], WT1 [122], SART-3 [123], and SOX11 [124, 125]. In general, immunotherapies consist of antibody-mediated immunotherapy, active immunotherapy that induces antitumor immunity in patients via a cancer vaccine, and adoptive or passive immunotherapy whereby tumor antigen-activated T cells are prepared ex vivo and administered to patients [114].

Tenascin is a well-known antigen associated with glioma and is an extracellular matrix molecule that is prominently expressed in the fibrillary matrix and perivascular patterns of gliomas [126, 127]. Multiple monoclonal antibodies (mAb) specific for human tenascin have also been generated [128, 129]. As EGFR is highly expressed by glioma cells, a chimeric mAb (cetuximab) has also been used in clinical trials but showed disappointing results [130]. More recently, a chimeric form of mAb ch806 administered to a patient with anaplastic astrocytoma showed good localization of the mAb at the tumor [131]. Various clinical trials have also studied the efficacy of various mAb to EGFR [114].

HER2-specific T cells against CD133+ cells generated by transduction with a retroviral vector encoding a HER2-specific chimeric antigen receptor have been used by investigators to show sustained regression of autologous GBM xenografts [132]. The same group also reported regression of experimental MB following transfer of HER2-specific T cells [133]. Similarly, IL-13 receptor alpha2 (IL13Ralpha2) is a glioma-restricted cell-surface epitope that is not otherwise detected within the CNS. Numerous preclinical studies have demonstrated the ability of L13-zetakine-redirected T cells to cause regression of GBM and GBM TSCs, as well as MB TSCs [134–137].

Dendritic cells (DCs) are the most potent antigen-presenting cells and have the ability to prime naïve T cells. A variety of tumor-associated antigens (specific tumor-associated peptides, tumor RNA and cDNA, tumor cell lysate, or apoptotic tumor cells) have been tested in numerous studies [138, 139]. Initial clinical trials using DC vaccines have shown to have strong systemic and intracranial T cell response and robust infiltration with T cells along with positive clinical outcomes [140–142]. Some studies have also suggested that eliminating the regulatory T cells would lead to improved anti-glioma immunity [143, 144].

TSC Differentiation

Due to their nature, TSCs play an important role in the tumorigenicity and maintenance of CNS tumors. While other agents aim to decrease the number of TSCs via specific targeting, differentiation agents aim to preferentially route the TSC into differentiating into progenitor cells. The strategy helps in decreasing the number of TSCs and gives rise to downstream tumor stem cells that are much more likely to be vulnerable to established therapeutics.

One of the first agents to be used as a differentiating agent for GBM TSCs was bone morphogenetic protein (BMP) 4 [145]. BMPs have an instructive role in the adult brain stem cell niche and favor the acquisition of an astroglial fate [146, 147]. Piccirillo and colleagues demonstrated that BMPs trigger the Smad signaling cascade in GBM cells. This was followed by a decrease in the size of CD133+ population and a decrease in their clonogenic ability [145].

A closer look at the oncogene BMI1 that regulates gene expression by modifying chromatin organization demonstrated that BMI1 was highly expressed in CD133+ cells. Knockdown of this gene using short hairpin RNA-expressing lentiviruses resulted in the inhibition of clonogenic potential in vitro and of brain tumor formation in vivo [148]. More recent research has shown the importance of BMI1 to self-renewal in CD133+ populations as well [149].

Metformin, a first-line drug for type II diabetes, was recently reported to possess anticancer properties affecting the survival TSCs in breast cancer models [150–152]. Würth and colleagues investigated the effect of metformin on glioma cells and reported a TSC-specific inhibition of Akt-dependent cell survival pathway that affected the self-renewal mechanisms [153]. Clinical trials using metformin for treatment of GBM are being conducted in the light of these promising results (NCT02149459 and NCT01430351).

Induction of autophagy has also shown to promote differentiation in glioma TSCs. Drugs such as rapamycin [154] and curcumin [155] trigger the differentiation cascade in TSCs by activating autophagy. Other differentiating targets include girdin, an actin-binding protein [156], and the vanilloid-2 cation channel [157]. Cannabinoids and sorafenib have also been documented to induce glioma TSC differentiation and deplete GBM TSCs [158].

Virotherapy and Gene Therapy

Among the emerging therapeutic options for CNS TSCs, virotherapy has shown noteworthy promise in terms of targeting glioma TSCs [159] (Fig. 3). Fueyo and colleagues constructed a tumor-selective adenovirus (Delta24) that carried a 24-bp deletion in the E1A region responsible for binding Rb protein. In vivo and in vitro results from their study demonstrated a potent lytic effect of glioma cells [160]. Later another group used a second-generation Delta24 (Delta24-hyCD) and

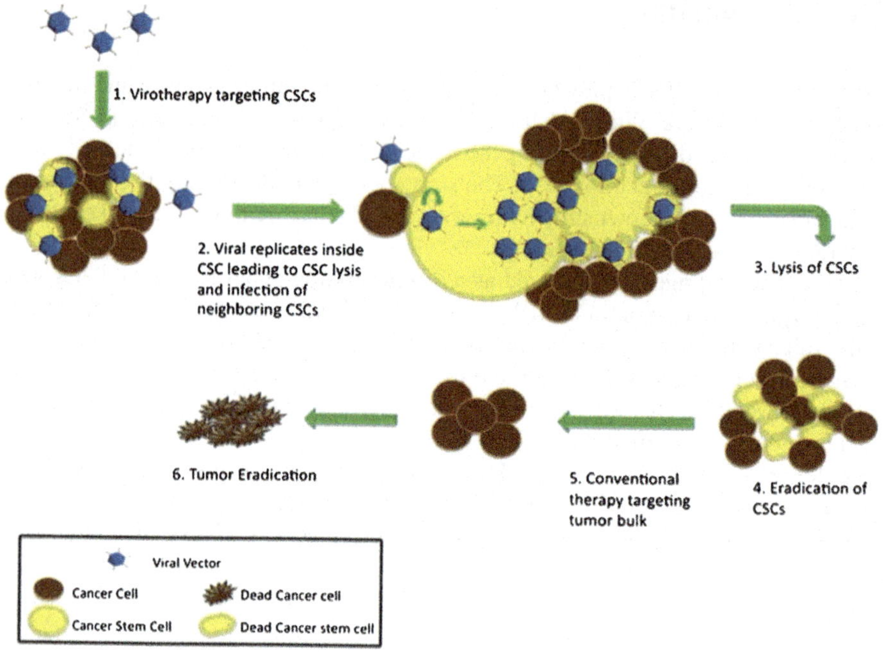

Fig. 3 Stem cell-targeted virotherapy. Adenoviral vectors are genetically modified to recognize and multiply only in cancer stem cells (CSCs). Viral replication in CSCs leads to destruction of CSCs and release of viral progeny, which in turn further infect neighboring stem cells. Repetition of this cycle leads to eradication of CSCs. Thus targeted therapy in addition to conventional therapy can lead to eradication of the tumor (reprinted with permission from Dey M et al. Stem Cell Rev. 2011 Mar;7(1):119–29. Cancer stem cells: the final frontier for glioma virotherapy)

exhibited significant chemosensitization and significant glioma control when 5-fluorocytosine was coupled with Delta24-hyCD [161].

In another study, a combination of adenoviral virotherapy and TMZ chemotherapy demonstrated a significant overexpression of autophagy markers, acidic vesicular organelles, and light-chain 3 (LC3) in vitro. In vivo studies showed significantly higher survival with combination therapy [162].

Gene silencing techniques can also be used to better understand the role of certain genes in the biology of TSCs and identify viable therapeutic targets. Bao and colleagues investigated the role of a neuronal cell adhesion molecule, L1CAM, in glioma TSCs using lentiviral mediated shRNA interference. They reported disrupted neurosphere formation, induced apoptosis, and inhibited growth of glioma TSCs [163]. Similarly, Wang and colleagues interrogated the significance of c-Myc expression in glioma TSCs using shRNA interference and showed that decreased expression of the target decreased proliferation and survival of TSCs [164].

Conclusion

Promising results from preclinical research using TSC-directed therapy have led to hopes for significant improvement in outcomes with high-grade CNS tumors. In this regard, the combination of conventional surgery, chemotherapy, and radiotherapy with TSC-targeted therapy may provide a new treatment approach to improve the response of CNS tumors. The potential efficacy of these therapeutic measures is being tested in various clinical trials and may direct future therapeutic interventions for CNS malignancies.

References

1. Stupp R, Mason WP, van den Bent MJ, Weller M, Fisher B, Taphoorn MJB, et al. Radiotherapy plus concomitant and adjuvant temozolomide for glioblastoma. N Engl J Med. 2005; 352(10):987–96.
2. Clarke MF, Dick JE, Dirks PB, Eaves CJ, Jamieson CHM, Jones DL, et al. Cancer stem cells–perspectives on current status and future directions: AACR Workshop on cancer stem cells. Cancer Res. 2006;66(19):9339–44.
3. Wakimoto H, Mohapatra G, Kanai R, Curry WT, Yip S, Nitta M, et al. Maintenance of primary tumor phenotype and genotype in glioblastoma stem cells. Neuro Oncol. 2012;14(2): 132–44.
4. Kelly JJP, Stechishin O, Chojnacki A, Lun X, Sun B, Senger DL, et al. Proliferation of human glioblastoma stem cells occurs independently of exogenous mitogens. Stem Cells (Dayton Ohio). 2009;27(8):1722–33.
5. Zhang J, Stevens MFG, Bradshaw TD. Temozolomide: mechanisms of action, repair and resistance. Curr Mol Pharmacol. 2012;5(1):102–14.
6. Hegi ME, Diserens A-C, Gorlia T, Hamou M-F, de Tribolet N, Weller M, et al. MGMT gene silencing and benefit from temozolomide in glioblastoma. N Engl J Med. 2005;352(10): 997–1003.
7. Cho D-Y, Lin S-Z, Yang W-K, Lee H-C, Hsu D-M, Lin H-L, et al. Targeting cancer stem cells for treatment of glioblastoma multiforme. Cell Transplant. 2013;22(4):731–9.
8. Liu G, Yuan X, Zeng Z, Tunici P, Ng H, Abdulkadir IR, et al. Analysis of gene expression and chemoresistance of CD133+ cancer stem cells in glioblastoma. Mol Cancer. 2006;5:67.
9. Fu J, Liu Z, Liu X, Chen F, Shi H, Pangjesse C, et al. Glioblastoma stem cells resistant to temozolomide-induced autophagy. Chin Med J (Engl). 2009;122(11):1255–9.
10. Nakai E, Park K, Yawata T, Chihara T, Kumazawa A, Nakabayashi H, et al. Enhanced MDR1 expression and chemoresistance of cancer stem cells derived from glioblastoma. Cancer Invest. 2009;27(9):901–8.
11. Beier D, Röhrl S, Pillai DR, Schwarz S, Kunz-Schughart LA, Leukel P, et al. Temozolomide preferentially depletes cancer stem cells in glioblastoma. Cancer Res. 2008;68(14): 5706–15.
12. Oka N, Soeda A, Noda S, Iwama T. Brain tumor stem cells from an adenoid glioblastoma multiforme. Neurol Med Chir (Tokyo). 2009;49(4):146–50. discussion 150–1.
13. Pérez Castillo A, Aguilar-Morante D, Morales-García JA, Dorado J. Cancer stem cells and brain tumors. Clin Transl Oncol Off Publ Fed Span Oncol Soc Natl Cancer Inst Mex. 2008;10(5):262–7.

14. Sakariassen PØ, Immervoll H, Chekenya M. Cancer stem cells as mediators of treatment resistance in brain tumors: status and controversies. Neoplasia. 2007;9(11):882–92.
15. Schmalz PGR, Shen MJ, Park JK. Treatment resistance mechanisms of malignant glioma tumor stem cells. Cancers. 2011;3(1):621–35.
16. Bao S, Wu Q, McLendon RE, Hao Y, Shi Q, Hjelmeland AB, et al. Glioma stem cells promote radioresistance by preferential activation of the DNA damage response. Nature. 2006; 444(7120):756–60.
17. Ropolo M, Daga A, Griffero F, Foresta M, Casartelli G, Zunino A, et al. Comparative analysis of DNA repair in stem and nonstem glioma cell cultures. Mol Cancer Res. 2009; 7(3):383–92.
18. Nicolis SK. Cancer stem cells and "stemness" genes in neuro-oncology. Neurobiol Dis. 2007;25(2):217–29.
19. Luther N, Cheung N-K, Souliopoulos EP, Karampelas I, Karempelas I, Bassiri D, et al. Interstitial infusion of glioma-targeted recombinant immunotoxin 8H9scFv-PE38. Mol Cancer Ther. 2010;9(4):1039–46.
20. Cho D-Y, Lin S-Z, Yang W-K, Hsu D-M, Lee H-C, Lee W-Y, et al. Recent advances of dendritic cells (DCs)-based immunotherapy for malignant gliomas. Cell Transplant. 2009; 18(9):977–83.
21. Piccirillo SGM, Vescovi AL. Brain tumour stem cells: possibilities of new therapeutic strategies. Expert Opin Biol Ther. 2007;7(8):1129–35.
22. Allenspach EJ, Maillard I, Aster JC, Pear WS. Notch signaling in cancer. Cancer Biol Ther. 2002;1(5):466–76.
23. Dang L, Fan X, Chaudhry A, Wang M, Gaiano N, Eberhart CG. Notch3 signaling initiates choroid plexus tumor formation. Oncogene. 2006;25(3):487–91.
24. Houde C, Li Y, Song L, Barton K, Zhang Q, Godwin J, et al. Overexpression of the NOTCH ligand JAG2 in malignant plasma cells from multiple myeloma patients and cell lines. Blood. 2004;104(12):3697–704.
25. Hovinga KE, Shimizu F, Wang R, Panagiotakos G, Van Der Heijden M, Moayedpardazi H, et al. Inhibition of notch signaling in glioblastoma targets cancer stem cells via an endothelial cell intermediate. Stem Cells (Dayton Ohio). 2010;28(6):1019–29.
26. Miyamoto Y, Maitra A, Ghosh B, Zechner U, Argani P, Iacobuzio-Donahue CA, et al. Notch mediates TGF alpha-induced changes in epithelial differentiation during pancreatic tumorigenesis. Cancer Cell. 2003;3(6):565–76.
27. Nickoloff BJ, Osborne BA, Miele L. Notch signaling as a therapeutic target in cancer: a new approach to the development of cell fate modifying agents. Oncogene. 2003;22(42): 6598–608.
28. Parr C, Watkins G, Jiang WG. The possible correlation of Notch-1 and Notch-2 with clinical outcome and tumour clinicopathological parameters in human breast cancer. Int J Mol Med. 2004;14(5):779–86.
29. Pece S, Serresi M, Santolini E, Capra M, Hulleman E, Galimberti V, et al. Loss of negative regulation by Numb over Notch is relevant to human breast carcinogenesis. J Cell Biol. 2004;167(2):215–21.
30. Purow BW, Haque RM, Noel MW, Su Q, Burdick MJ, Lee J, et al. Expression of Notch-1 and its ligands, Delta-like-1 and Jagged-1, is critical for glioma cell survival and proliferation. Cancer Res. 2005;65(6):2353–63.
31. Zhu TS, Costello MA, Talsma CE, Flack CG, Crowley JG, Hamm LL, et al. Endothelial cells create a stem cell niche in glioblastoma by providing NOTCH ligands that nurture self-renewal of cancer stem-like cells. Cancer Res. 2011;71(18):6061–72.
32. Galli R, Binda E, Orfanelli U, Cipelletti B, Gritti A, De Vitis S, et al. Isolation and characterization of tumorigenic, stem-like neural precursors from human glioblastoma. Cancer Res. 2004;64(19):7011–21.
33. Fan X, Khaki L, Zhu TS, Soules ME, Talsma CE, Gul N, et al. NOTCH pathway blockade depletes CD133-positive glioblastoma cells and inhibits growth of tumor neurospheres and xenografts. Stem Cells (Dayton Ohio). 2010;28(1):5–16.

34. Dahmane N, Ruiz i Altaba A. Sonic hedgehog regulates the growth and patterning of the cerebellum. Dev Camb Engl. 1999;126(14):3089–100.
35. Wechsler-Reya RJ, Scott MP. Control of neuronal precursor proliferation in the cerebellum by Sonic Hedgehog. Neuron. 1999;22(1):103–14.
36. Lee A, Kessler JD, Read T-A, Kaiser C, Corbeil D, Huttner WB, et al. Isolation of neural stem cells from the postnatal cerebellum. Nat Neurosci. 2005;8(6):723–9.
37. Michael LE, Westerman BA, Ermilov AN, Wang A, Ferris J, Liu J, et al. Bmi1 is required for Hedgehog pathway-driven medulloblastoma expansion. Neoplasia. 2008;10(12):1343–9. 5p following 1349.
38. Wang X, Venugopal C, Manoranjan B, McFarlane N, O'Farrell E, Nolte S, et al. Sonic hedgehog regulates Bmi1 in human medulloblastoma brain tumor-initiating cells. Oncogene. 2012; 31(2):187–99.
39. Kinzler KW, Bigner SH, Bigner DD, Trent JM, Law ML, O'Brien SJ, et al. Identification of an amplified, highly expressed gene in a human glioma. Science. 1987;236(4797):70–3.
40. Bar EE, Chaudhry A, Lin A, Fan X, Schreck K, Matsui W, et al. Cyclopamine-mediated hedgehog pathway inhibition depletes stem-like cancer cells in glioblastoma. Stem Cells (Dayton Ohio). 2007;25(10):2524–33.
41. Clement V, Sanchez P, de Tribolet N, Radovanovic I, Ruiz i Altaba A. HEDGEHOG-GLI1 signaling regulates human glioma growth, cancer stem cell self-renewal, and tumorigenicity. Curr Biol. 2007;17(2):165–72.
42. Dixit D, Ghildiyal R, Anto NP, Ghosh S, Sharma V, Sen E. Guggulsterone sensitizes glioblastoma cells to Sonic hedgehog inhibitor SANT-1 induced apoptosis in a Ras/NFκB dependent manner. Cancer Lett. 2013;336(2):347–58.
43. Barami K. Relationship of neural stem cells with their vascular niche: implications in the malignant progression of gliomas. J Clin Neurosci. 2008;15(11):1193–7.
44. Yan G-N, Lv Y-F, Yang L, Yao X-H, Cui Y-H, Guo D-Y. Glioma stem cells enhance endothelial cell migration and proliferation via the Hedgehog pathway. Oncol Lett. 2013;6(5):1524–30.
45. Meiss F, Zeiser R. Vismodegib. Recent Results Cancer Res Fortschritte Krebsforsch Prog Dans Rech Sur Cancer. 2014;201:405–17.
46. Rudin CM, Hann CL, Laterra J, Yauch RL, Callahan CA, Fu L, et al. Treatment of medulloblastoma with hedgehog pathway inhibitor GDC-0449. N Engl J Med. 2009;361(12):1173–8.
47. Bao S, Wu Q, Sathornsumetee S, Hao Y, Li Z, Hjelmeland AB, et al. Stem cell-like glioma cells promote tumor angiogenesis through vascular endothelial growth factor. Cancer Res. 2006;66(16):7843–8.
48. Calabrese C, Poppleton H, Kocak M, Hogg TL, Fuller C, Hamner B, et al. A perivascular niche for brain tumor stem cells. Cancer Cell. 2007;11(1):69–82.
49. Gilbert MR, Dignam JJ, Armstrong TS, Wefel JS, Blumenthal DT, Vogelbaum MA, et al. A randomized trial of bevacizumab for newly diagnosed glioblastoma. N Engl J Med. 2014;370(8):699–708.
50. Batchelor TT, Mulholland P, Neyns B, Nabors LB, Campone M, Wick A, et al. Phase III randomized trial comparing the efficacy of cediranib as monotherapy, and in combination with lomustine, versus lomustine alone in patients with recurrent glioblastoma. J Clin Oncol. 2013;31(26):3212–8.
51. Cancer Genome Atlas Research Network. Comprehensive genomic characterization defines human glioblastoma genes and core pathways. Nature. 2008;455(7216):1061–8.
52. Parsons DW, Jones S, Zhang X, Lin JC-H, Leary RJ, Angenendt P, et al. An integrated genomic analysis of human glioblastoma multiforme. Science. 2008;321(5897):1807–12.
53. Inda M-M, Bonavia R, Mukasa A, Narita Y, Sah DWY, Vandenberg S, et al. Tumor heterogeneity is an active process maintained by a mutant EGFR-induced cytokine circuit in glioblastoma. Genes Dev. 2010;24(16):1731–45.
54. Jin X, Yin J, Kim S-H, Sohn Y-W, Beck S, Lim YC, et al. EGFR-AKT-Smad signaling promotes formation of glioma stem-like cells and tumor angiogenesis by ID3-driven cytokine induction. Cancer Res. 2011;71(22):7125–34.

55. Feng H, Hu B, Jarzynka MJ, Li Y, Keezer S, Johns TG, et al. Phosphorylation of dedicator of cytokinesis 1 (Dock180) at tyrosine residue Y722 by Src family kinases mediates EGFRvIII-driven glioblastoma tumorigenesis. Proc Natl Acad Sci U S A. 2012;109(8):3018–23.
56. Feng H, Hu B, Vuori K, Sarkaria JN, Furnari FB, Cavenee WK, et al. EGFRvIII stimulates glioma growth and invasion through PKA-dependent serine phosphorylation of Dock180. Oncogene. 2014;33(19):2504–12.
57. Raizer JJ, Abrey LE, Lassman AB, Chang SM, Lamborn KR, Kuhn JG, et al. A phase II trial of erlotinib in patients with recurrent malignant gliomas and nonprogressive glioblastoma multiforme postradiation therapy. Neuro Oncol. 2010;12(1):95–103.
58. Hasselbalch B, Eriksen JG, Broholm H, Christensen IJ, Grunnet K, Horsman MR, et al. Prospective evaluation of angiogenic, hypoxic and EGFR-related biomarkers in recurrent glioblastoma multiforme treated with cetuximab, bevacizumab and irinotecan. APMIS Acta Pathol Microbiol Immunol Scand. 2010;118(8):585–94.
59. Hegi ME, Diserens A-C, Bady P, Kamoshima Y, Kouwenhoven MCM, Delorenzi M, et al. Pathway analysis of glioblastoma tissue after preoperative treatment with the EGFR tyrosine kinase inhibitor gefitinib—a phase II trial. Mol Cancer Ther. 2011;10(6):1102–12.
60. Lindvall C, Bu W, Williams BO, Li Y. Wnt signaling, stem cells, and the cellular origin of breast cancer. Stem Cell Rev. 2007;3(2):157–68.
61. Lustig B, Behrens J. The Wnt signaling pathway and its role in tumor development. J Cancer Res Clin Oncol. 2003;129(4):199–221.
62. Polakis P. The many ways of Wnt in cancer. Curr Opin Genet Dev. 2007;17(1):45–51.
63. Augustin I, Goidts V, Bongers A, Kerr G, Vollert G, Radlwimmer B, et al. The Wnt secretion protein Evi/Gpr177 promotes glioma tumourigenesis. EMBO Mol Med. 2012;4(1):38–51.
64. Sandberg CJ, Altschuler G, Jeong J, Strømme KK, Stangeland B, Murrell W, et al. Comparison of glioma stem cells to neural stem cells from the adult human brain identifies dysregulated Wnt-signaling and a fingerprint associated with clinical outcome. Exp Cell Res. 2013;319(14):2230–43.
65. Kim KH, Seol HJ, Kim EH, Rheey J, Jin HJ, Lee Y, et al. Wnt/β-catenin signaling is a key downstream mediator of MET signaling in glioblastoma stem cells. Neuro Oncol. 2013; 15(2):161–71.
66. Rogers HA, Sousa S, Salto C, Arenas E, Coyle B, Grundy RG. WNT/β-catenin pathway activation in Myc immortalised cerebellar progenitor cells inhibits neuronal differentiation and generates tumours resembling medulloblastoma. Br J Cancer. 2012;107(7):1144–52.
67. Mascaro Cordeiro B, Dias Oliveira I, de Seixas Alves MT, Saba-Silva N, Capellano AM, Cavalheiro S, et al. SHH, WNT, and NOTCH pathways in medulloblastoma: when cancer stem cells maintain self-renewal and differentiation properties. Childs Nerv Syst. 2014;30(7):1165–72.
68. Di Cristofano A, Pandolfi PP. The multiple roles of PTEN in tumor suppression. Cell. 2000;100(4):387–90.
69. Marian CO, Cho SK, McEllin BM, Maher EA, Hatanpaa KJ, Madden CJ, et al. The telomerase antagonist, imetelstat, efficiently targets glioblastoma tumor-initiating cells leading to decreased proliferation and tumor growth. Clin Cancer Res. 2010;16(1):154–63.
70. Michelakis ED, Sutendra G, Dromparis P, Webster L, Haromy A, Niven E, et al. Metabolic modulation of glioblastoma with dichloroacetate. Sci Transl Med. 2010;2(31):31ra34.
71. Colen CB, Shen Y, Ghoddoussi F, Yu P, Francis TB, Koch BJ, et al. Metabolic targeting of lactate efflux by malignant glioma inhibits invasiveness and induces necrosis: an in vivo study. Neoplasia. 2011;13(7):620–32.
72. Chen L, Zhang R, Li P, Liu Y, Qin K, Fa Z-Q, et al. P53-induced microRNA-107 inhibits proliferation of glioma cells and down-regulates the expression of CDK6 and Notch-2. Neurosci Lett. 2013;534:327–32.
73. Guessous F, Zhang Y, Kofman A, Catania A, Li Y, Schiff D, et al. microRNA-34a is tumor suppressive in brain tumors and glioma stem cells. Cell Cycle Georget Tex. 2010;9(6): 1031–6.

74. Li Y, Guessous F, Zhang Y, Dipierro C, Kefas B, Johnson E, et al. MicroRNA-34a inhibits glioblastoma growth by targeting multiple oncogenes. Cancer Res. 2009;69(19):7569–76.
75. Hirose Y, Berger MS, Pieper RO. Abrogation of the Chk1-mediated G(2) checkpoint pathway potentiates temozolomide-induced toxicity in a p53-independent manner in human glioblastoma cells. Cancer Res. 2001;61(15):5843–9.
76. Hirose Y, Katayama M, Stokoe D, Haas-Kogan DA, Berger MS, Pieper RO. The p38 mitogen-activated protein kinase pathway links the DNA mismatch repair system to the G2 checkpoint and to resistance to chemotherapeutic DNA-methylating agents. Mol Cell Biol. 2003;23(22):8306–15.
77. Solit DB, Basso AD, Olshen AB, Scher HI, Rosen N. Inhibition of heat shock protein 90 function down-regulates Akt kinase and sensitizes tumors to Taxol. Cancer Res. 2003; 63(9):2139–44.
78. Hirose Y, Kreklau EL, Erickson LC, Berger MS, Pieper RO. Delayed repletion of O6-methylguanine-DNA methyltransferase resulting in failure to protect the human glioblastoma cell line SF767 from temozolomide-induced cytotoxicity. J Neurosurg. 2003;98(3): 591–8.
79. Ferrarini M, Heltai S, Zocchi MR, Rugarli C. Unusual expression and localization of heat-shock proteins in human tumor cells. Int J Cancer. 1992;51(4):613–9.
80. Ohba S, Hirose Y, Yoshida K, Yazaki T, Kawase T. Inhibition of 90-kD heat shock protein potentiates the cytotoxicity of chemotherapeutic agents in human glioma cells. J Neurosurg. 2010;112(1):33–42.
81. Sauvageot CM-E, Weatherbee JL, Kesari S, Winters SE, Barnes J, Dellagatta J, et al. Efficacy of the HSP90 inhibitor 17-AAG in human glioma cell lines and tumorigenic glioma stem cells. Neuro Oncol. 2009;11(2):109–21.
82. Tentori L, Leonetti C, Scarsella M, D'Amati G, Vergati M, Portarena I, et al. Systemic administration of GPI 15427, a novel poly(ADP-ribose) polymerase-1 inhibitor, increases the antitumor activity of temozolomide against intracranial melanoma, glioma, lymphoma. Clin Cancer Res. 2003;9(14):5370–9.
83. Tentori L, Leonetti C, Scarsella M, Vergati M, Xu W, Calvin D, et al. Brain distribution and efficacy as chemosensitizer of an oral formulation of PARP-1 inhibitor GPI 15427 in experimental models of CNS tumors. Int J Oncol. 2005;26(2):415–22.
84. Tentori L, Leonetti C, Scarsella M, Muzi A, Vergati M, Forini O, et al. Poly(ADP-ribose) glycohydrolase inhibitor as chemosensitiser of malignant melanoma for temozolomide. Eur J Cancer Oxf Engl 1990. 2005;41(18):2948–57.
85. Warrier S, Balu SK, Kumar AP, Millward M, Dharmarajan A. Wnt antagonist, secreted frizzled-related protein 4 (sFRP4), increases chemotherapeutic response of glioma stem-like cells. Oncol Res. 2014;21(2):93–102.
86. Bota DA, Alexandru D, Keir ST, Bigner D, Vredenburgh J, Friedman HS. Proteasome inhibition with bortezomib induces cell death in GBM stem-like cells and temozolomide-resistant glioma cell lines, but stimulates GBM stem-like cells' VEGF production and angiogenesis. J Neurosurg. 2013;119(6):1415–23.
87. Xu Y, Stamenkovic I, Yu Q. CD44 attenuates activation of the hippo signaling pathway and is a prime therapeutic target for glioblastoma. Cancer Res. 2010;70(6):2455–64.
88. Wachsberger PR, Lawrence RY, Liu Y, Xia X, Andersen B, Dicker AP. Cediranib enhances control of wild type EGFR and EGFRvIII-expressing gliomas through potentiating temozolomide, but not through radiosensitization: implications for the clinic. J Neurooncol. 2011;105(2):181–90.
89. Kang M-K, Kang S-K. Pharmacologic blockade of chloride channel synergistically enhances apoptosis of chemotherapeutic drug-resistant cancer stem cells. Biochem Biophys Res Commun. 2008;373(4):539–44.
90. Folkins C, Man S, Xu P, Shaked Y, Hicklin DJ, Kerbel RS. Anticancer therapies combining antiangiogenic and tumor cell cytotoxic effects reduce the tumor stem-like cell fraction in glioma xenograft tumors. Cancer Res. 2007;67(8):3560–4.

91. Zhang M, Kleber S, Röhrich M, Timke C, Han N, Tuettenberg J, et al. Blockade of TGF-β signaling by the TGFβR-I kinase inhibitor LY2109761 enhances radiation response and prolongs survival in glioblastoma. Cancer Res. 2011;71(23):7155–67.
92. Hardee ME, Marciscano AE, Medina-Ramirez CM, Zagzag D, Narayana A, Lonning SM, et al. Resistance of glioblastoma-initiating cells to radiation mediated by the tumor microenvironment can be abolished by inhibiting transforming growth factor-β. Cancer Res. 2012;72(16):4119–29.
93. Geoerger B, Gaspar N, Opolon P, Morizet J, Devanz P, Lecluse Y, et al. EGFR tyrosine kinase inhibition radiosensitizes and induces apoptosis in malignant glioma and childhood ependymoma xenografts. Int J Cancer. 2008;123(1):209–16.
94. Kang KB, Zhu C, Wong YL, Gao Q, Ty A, Wong MC. Gefitinib radiosensitizes stem-like glioma cells: inhibition of epidermal growth factor receptor-Akt-DNA-PK signaling, accompanied by inhibition of DNA double-strand break repair. Int J Radiat Oncol Biol Phys. 2012;83(1):e43–52.
95. Stea B, Falsey R, Kislin K, Patel J, Glanzberg H, Carey S, et al. Time and dose-dependent radiosensitization of the glioblastoma multiforme U251 cells by the EGF receptor tyrosine kinase inhibitor ZD1839 ('Iressa'). Cancer Lett. 2003;202(1):43–51.
96. Chakravarti A, Wang M, Robins HI, Lautenschlaeger T, Curran WJ, Brachman DG, et al. RTOG 0211: a phase 1/2 study of radiation therapy with concurrent gefitinib for newly diagnosed glioblastoma patients. Int J Radiat Oncol Biol Phys. 2013;85(5):1206–11.
97. Drappatz J, Norden AD, Wong ET, Doherty LM, Lafrankie DC, Ciampa A, et al. Phase I study of vandetanib with radiotherapy and temozolomide for newly diagnosed glioblastoma. Int J Radiat Oncol Biol Phys. 2010;78(1):85–90.
98. Ashizawa T, Miyata H, Iizuka A, Komiyama M, Oshita C, Kume A, et al. Effect of the STAT3 inhibitor STX-0119 on the proliferation of cancer stem-like cells derived from recurrent glioblastoma. Int J Oncol. 2013;43(1):219–27.
99. Stechishin OD, Luchman HA, Ruan Y, Blough MD, Nguyen SA, Kelly JJ, et al. On-target JAK2/STAT3 inhibition slows disease progression in orthotopic xenografts of human glioblastoma brain tumor stem cells. Neuro Oncol. 2013;15(2):198–207.
100. Yang Y-P, Chang Y-L, Huang P-I, Chiou G-Y, Tseng L-M, Chiou S-H, et al. Resveratrol suppresses tumorigenicity and enhances radiosensitivity in primary glioblastoma tumor initiating cells by inhibiting the STAT3 axis. J Cell Physiol. 2012;227(3):976–93.
101. Chang C-J, Chiang C-H, Song W-S, Tsai S-K, Woung L-C, Chang C-H, et al. Inhibition of phosphorylated STAT3 by cucurbitacin I enhances chemoradiosensitivity in medulloblastoma-derived cancer stem cells. Childs Nerv Syst. 2012;28(3):363–73.
102. Liu D, Hu G, Long G, Qiu H, Mei Q, Hu G. Celecoxib induces apoptosis and cell-cycle arrest in nasopharyngeal carcinoma cell lines via inhibition of STAT3 phosphorylation. Acta Pharmacol Sin. 2012;33(5):682–90.
103. Reed S, Li H, Li C, Lin J. Celecoxib inhibits STAT3 phosphorylation and suppresses cell migration and colony forming ability in rhabdomyosarcoma cells. Biochem Biophys Res Commun. 2011;407(3):450–5.
104. Yang M-Y, Lee H-T, Chen C-M, Shen C-C, Ma H-I. Celecoxib suppresses the phosphorylation of STAT3 protein and can enhance the radiosensitivity of medulloblastoma-derived cancer stem-like cells. Int J Mol Sci. 2014;15(6):11013–29.
105. Göttlicher M, Minucci S, Zhu P, Krämer OH, Schimpf A, Giavara S, et al. Valproic acid defines a novel class of HDAC inhibitors inducing differentiation of transformed cells. EMBO J. 2001;20(24):6969–78.
106. Kostrouchová M, Kostrouch Z, Kostrouchová M. Valproic acid, a molecular lead to multiple regulatory pathways. Folia Biol (Praha). 2007;53(2):37–49.
107. Phiel CJ, Zhang F, Huang EY, Guenther MG, Lazar MA, Klein PS. Histone deacetylase is a direct target of valproic acid, a potent anticonvulsant, mood stabilizer, and teratogen. J Biol Chem. 2001;276(39):36734–41.

108. Chen C-H, Chang Y-J, Ku MSB, Chung K-T, Yang J-T. Enhancement of temozolomide-induced apoptosis by valproic acid in human glioma cell lines through redox regulation. J Mol Med (Berl). 2011;89(3):303–15.
109. Van Nifterik KA, Van den Berg J, Slotman BJ, Lafleur MVM, Sminia P, Stalpers LJA. Valproic acid sensitizes human glioma cells for temozolomide and γ-radiation. J Neurooncol. 2012; 107(1):61–7.
110. Mukherjee B, Tomimatsu N, Amancherla K, Camacho CV, Pichamoorthy N, Burma S. The dual PI3K/mTOR inhibitor NVP-BEZ235 is a potent inhibitor of ATM- and DNA-PKCs-mediated DNA damage responses. Neoplasia. 2012;14(1):34–43.
111. Gil del Alcazar CR, Hardebeck MC, Mukherjee B, Tomimatsu N, Gao X, Yan J, et al. Inhibition of DNA double-strand break repair by the dual PI3K/mTOR inhibitor NVP-BEZ235 as a strategy for radiosensitization of glioblastoma. Clin Cancer Res. 2014;20(5):1235–48.
112. Wang W, Long L, Yang N, Zhang Q, Ji W, Zhao J, et al. NVP-BEZ235, a novel dual PI3K/mTOR inhibitor, enhances the radiosensitivity of human glioma stem cells in vitro. Acta Pharmacol Sin. 2013;34(5):681–90.
113. Albesiano E, Han JE, Lim M. Mechanisms of local immunoresistance in glioma. Neurosurg Clin N Am. 2010;21(1):17–29.
114. Toda M. Glioma stem cells and immunotherapy for the treatment of malignant gliomas. ISRN Oncol. 2013;2013:673793.
115. Facoetti A, Nano R, Zelini P, Morbini P, Benericetti E, Ceroni M, et al. Human leukocyte antigen and antigen processing machinery component defects in astrocytic tumors. Clin Cancer Res. 2005;11(23):8304–11.
116. Kuramoto T. Detection of MAGE-1 tumor antigen in brain tumor. Kurume Med J. 1997; 44(1):43–51.
117. Ueda R, Yoshida K, Kawakami Y, Kawase T, Toda M. Expression of a transcriptional factor, SOX6, in human gliomas. Brain Tumor Pathol. 2004;21(1):35–8.
118. Liu G, Ying H, Zeng G, Wheeler CJ, Black KL, Yu JS. HER-2, gp100, and MAGE-1 are expressed in human glioblastoma and recognized by cytotoxic T cells. Cancer Res. 2004; 64(14):4980–6.
119. Chi DD, Merchant RE, Rand R, Conrad AJ, Garrison D, Turner R, et al. Molecular detection of tumor-associated antigens shared by human cutaneous melanomas and gliomas. Am J Pathol. 1997;150(6):2143–52.
120. Heimberger AB, Crotty LE, Archer GE, Hess KR, Wikstrand CJ, Friedman AH, et al. Epidermal growth factor receptor VIII peptide vaccination is efficacious against established intracerebral tumors. Clin Cancer Res. 2003;9(11):4247–54.
121. Okano F, Storkus WJ, Chambers WH, Pollack IF, Okada H. Identification of a novel HLA-A*0201-restricted, cytotoxic T lymphocyte epitope in a human glioma-associated antigen, interleukin 13 receptor alpha2 chain. Clin Cancer Res. 2002;8(9):2851–5.
122. Hashiba T, Izumoto S, Kagawa N, Suzuki T, Hashimoto N, Maruno M, et al. Expression of WT1 protein and correlation with cellular proliferation in glial tumors. Neurol Med Chir (Tokyo). 2007;47(4):165–70. discussion 170.
123. Murayama K, Kobayashi T, Imaizumi T, Matsunaga K, Kuramoto T, Shigemori M, et al. Expression of the SART3 tumor-rejection antigen in brain tumors and induction of cytotoxic T lymphocytes by its peptides. J Immunother Hagerstown Md 1997. 2000;23(5):511–8.
124. Schmitz M, Wehner R, Stevanovic S, Kiessling A, Rieger MA, Temme A, et al. Identification of a naturally processed T cell epitope derived from the glioma-associated protein SOX11. Cancer Lett. 2007;245(1–2):331–6.
125. Cheever MA, Allison JP, Ferris AS, Finn OJ, Hastings BM, Hecht TT, et al. The prioritization of cancer antigens: a national cancer institute pilot project for the acceleration of translational research. Clin Cancer Res. 2009;15(17):5323–37.
126. Kurpad SN, Zhao XG, Wikstrand CJ, Batra SK, McLendon RE, Bigner DD. Tumor antigens in astrocytic gliomas. Glia. 1995;15(3):244–56.

127. Bigner DD, Brown M, Coleman RE, Friedman AH, Friedman HS, McLendon RE, et al. Phase I studies of treatment of malignant gliomas and neoplastic meningitis with 131I-radiolabeled monoclonal antibodies anti-tenascin 81C6 and anti-chondroitin proteoglycan sulfate Me1-14F (ab')2—a preliminary report. J Neurooncol. 1995;24(1):109–22.
128. Riva P, Arista A, Franceschi G, Frattarelli M, Sturiale C, Riva N, et al. Local treatment of malignant gliomas by direct infusion of specific monoclonal antibodies labeled with 131I: comparison of the results obtained in recurrent and newly diagnosed tumors. Cancer Res. 1995;55(23 Suppl):5952s–6s.
129. Murphy-Ullrich JE, Lightner VA, Aukhil I, Yan YZ, Erickson HP, Höök M. Focal adhesion integrity is downregulated by the alternatively spliced domain of human tenascin. J Cell Biol. 1991;115(4):1127–36.
130. Neyns B, Sadones J, Joosens E, Bouttens F, Verbeke L, Baurain J-F, et al. Stratified phase II trial of cetuximab in patients with recurrent high-grade glioma. Ann Oncol. 2009;20(9):1596–603.
131. Scott AM, Lee F-T, Tebbutt N, Herbertson R, Gill SS, Liu Z, et al. A phase I clinical trial with monoclonal antibody ch806 targeting transitional state and mutant epidermal growth factor receptors. Proc Natl Acad Sci U S A. 2007;104(10):4071–6.
132. Ahmed N, Salsman VS, Kew Y, Shaffer D, Powell S, Zhang YJ, et al. HER2-specific T cells target primary glioblastoma stem cells and induce regression of autologous experimental tumors. Clin Cancer Res. 2010;16(2):474–85.
133. Ahmed N, Ratnayake M, Savoldo B, Perlaky L, Dotti G, Wels WS, et al. Regression of experimental medulloblastoma following transfer of HER2-specific T cells. Cancer Res. 2007;67(12):5957–64.
134. Kahlon KS, Brown C, Cooper LJN, Raubitschek A, Forman SJ, Jensen MC. Specific recognition and killing of glioblastoma multiforme by interleukin 13-zetakine redirected cytolytic T cells. Cancer Res. 2004;64(24):9160–6.
135. Kong S, Sengupta S, Tyler B, Bais AJ, Ma Q, Doucette S, et al. Suppression of human glioma xenografts with second-generation IL13R-specific chimeric antigen receptor-modified T cells. Clin Cancer Res. 2012;18(21):5949–60.
136. Brown CE, Starr R, Aguilar B, Shami AF, Martinez C, D'Apuzzo M, et al. Stem-like tumor-initiating cells isolated from IL13Rα2 expressing gliomas are targeted and killed by IL13-zetakine-redirected T Cells. Clin Cancer Res. 2012;18(8):2199–209.
137. Stastny MJ, Brown CE, Ruel C, Jensen MC. Medulloblastomas expressing IL13Ralpha2 are targets for IL13-zetakine+cytolytic T cells. J Pediatr Hematol Oncol. 2007;29(10):669–77.
138. Soling A, Rainov NG. Dendritic cell therapy of primary brain tumors. Mol Med (Camb Mass). 2001;7(10):659–67.
139. Ji J, Black KL, Yu JS. Glioma stem cell research for the development of immunotherapy. Neurosurg Clin N Am. 2010;21(1):159–66.
140. Yu JS, Wheeler CJ, Zeltzer PM, Ying H, Finger DN, Lee PK, et al. Vaccination of malignant glioma patients with peptide-pulsed dendritic cells elicits systemic cytotoxicity and intracranial T-cell infiltration. Cancer Res. 2001;61(3):842–7.
141. Yu JS, Liu G, Ying H, Yong WH, Black KL, Wheeler CJ. Vaccination with tumor lysate-pulsed dendritic cells elicits antigen-specific, cytotoxic T-cells in patients with malignant glioma. Cancer Res. 2004;64(14):4973–9.
142. Liau LM, Prins RM, Kiertscher SM, Odesa SK, Kremen TJ, Giovannone AJ, et al. Dendritic cell vaccination in glioblastoma patients induces systemic and intracranial T-cell responses modulated by the local central nervous system tumor microenvironment. Clin Cancer Res. 2005;11(15):5515–25.
143. Grauer OM, Sutmuller RPM, van Maren W, Jacobs JFM, Bennink E, Toonen LWJ, et al. Elimination of regulatory T cells is essential for an effective vaccination with tumor lysate-pulsed dendritic cells in a murine glioma model. Int J Cancer. 2008;122(8):1794–802.
144. Maes W, Rosas GG, Verbinnen B, Boon L, De Vleeschouwer S, Ceuppens JL, et al. DC vaccination with anti-CD25 treatment leads to long-term immunity against experimental glioma. Neuro Oncol. 2009;11(5):529–42.

145. Piccirillo SGM, Reynolds BA, Zanetti N, Lamorte G, Binda E, Broggi G, et al. Bone morphogenetic proteins inhibit the tumorigenic potential of human brain tumour-initiating cells. Nature. 2006;444(7120):761–5.
146. Lim DA, Tramontin AD, Trevejo JM, Herrera DG, García-Verdugo JM, Alvarez-Buylla A. Noggin antagonizes BMP signaling to create a niche for adult neurogenesis. Neuron. 2000;28(3):713–26.
147. Panchision DM, McKay RDG. The control of neural stem cells by morphogenic signals. Curr Opin Genet Dev. 2002;12(4):478–87.
148. Abdouh M, Facchino S, Chatoo W, Balasingam V, Ferreira J, Bernier G. BMI1 sustains human glioblastoma multiforme stem cell renewal. J Neurosci. 2009;29(28):8884–96.
149. Venugopal C, Li N, Wang X, Manoranjan B, Hawkins C, Gunnarsson T, et al. Bmi1 marks intermediate precursors during differentiation of human brain tumor initiating cells. Stem Cell Res. 2012;8(2):141–53.
150. Hirsch HA, Iliopoulos D, Tsichlis PN, Struhl K. Metformin selectively targets cancer stem cells, and acts together with chemotherapy to block tumor growth and prolong remission. Cancer Res. 2009;69(19):7507–11.
151. Jung J-W, Park S-B, Lee S-J, Seo M-S, Trosko JE, Kang K-S. Metformin represses self-renewal of the human breast carcinoma stem cells via inhibition of estrogen receptor-mediated OCT4 expression. PLoS One. 2011;6(11):e28068.
152. Song CW, Lee H, Dings RPM, Williams B, Powers J, Santos TD, et al. Metformin kills and radiosensitizes cancer cells and preferentially kills cancer stem cells. Sci Rep. 2012;2:362.
153. Wurth R, Pattarozzi A, Gatti M, Bajetto A, Corsaro A, Parodi A, et al. Metformin selectively affects human glioblastoma tumor-initiating cell viability. Cell Cycle. 2013;12(1):145–56.
154. Zhuang W, Li B, Long L, Chen L, Huang Q, Liang Z. Induction of autophagy promotes differentiation of glioma-initiating cells and their radiosensitivity. Int J Cancer. 2011;129(11):2720–31.
155. Zhuang W, Long L, Zheng B, Ji W, Yang N, Zhang Q, et al. Curcumin promotes differentiation of glioma-initiating cells by inducing autophagy. Cancer Sci. 2012;103(4):684–90.
156. Natsume A, Kato T, Kinjo S, Enomoto A, Toda H, Shimato S, et al. Girdin maintains the stemness of glioblastoma stem cells. Oncogene. 2012;31(22):2715–24.
157. Morelli MB, Nabissi M, Amantini C, Farfariello V, Ricci-Vitiani L, di Martino S, et al. The transient receptor potential vanilloid-2 cation channel impairs glioblastoma stem-like cell proliferation and promotes differentiation. Int J Cancer. 2012;131(7):E1067–77.
158. Aguado T, Carracedo A, Julien B, Velasco G, Milman G, Mechoulam R, et al. Cannabinoids induce glioma stem-like cell differentiation and inhibit gliomagenesis. J Biol Chem. 2007;282(9):6854–62.
159. Dey M, Ulasov IV, Tyler MA, Sonabend AM, Lesniak MS. Cancer stem cells: the final frontier for glioma virotherapy. Stem Cell Rev. 2011;7(1):119–29.
160. Fueyo J, Gomez-Manzano C, Alemany R, Lee PS, McDonnell TJ, Mitlianga P, et al. A mutant oncolytic adenovirus targeting the Rb pathway produces anti-glioma effect in vivo. Oncogene. 2000;19(1):2–12.
161. Conrad C, Miller CR, Ji Y, Gomez-Manzano C, Bharara S, McMurray JS, et al. Delta24-hyCD adenovirus suppresses glioma growth in vivo by combining oncolysis and chemosensitization. Cancer Gene Ther. 2005;12(3):284–94.
162. Ulasov IV, Sonabend AM, Nandi S, Khramtsov A, Han Y, Lesniak MS. Combination of adenoviral virotherapy and temozolomide chemotherapy eradicates malignant glioma through autophagic and apoptotic cell death in vivo. Br J Cancer. 2009;100(7):1154–64.
163. Bao S, Wu Q, Li Z, Sathornsumetee S, Wang H, McLendon RE, et al. Targeting cancer stem cells through L1CAM suppresses glioma growth. Cancer Res. 2008;68(15):6043–8.
164. Wang J, Wang H, Li Z, Wu Q, Lathia JD, McLendon RE, et al. c-Myc is required for maintenance of glioma cancer stem cells. PLoS One. 2008;3(11):e3769.

Index

M. Ehtesham (ed.), *Stem Cell Biology in Neoplasms of the Central Nervous System*, Advances in Experimental Medicine and Biology 853,
DOI 10.1007/978-3-319-16537-0

CPSIA information can be obtained at www.ICGtesting.com
Printed in the USA
LVOW02*1658020615

440877LV00001B/44/P

9 783319 165363